SOUTH ASIAN INSECURITY AND THE GREAT POWERS

Also by Barry Buzan

PEOPLE, STATES AND FEAR

SEABED POLITICS

CHANGE AND THE STUDY OF INTERNATIONAL RELATIONS
(*editor with R. J. Barry Jones*)

Also by Gowher Rizvi

LORD LINLITHGOW AND INDIA, 1936–43

PERSPECTIVES ON IMPERIALISM AND DECOLONIZATION
(*editor with R. Holland*)

INDO–BRITISH RELATIONS IN RETROSPECT
(*editor with A. Copley*)

SOUTH ASIAN INSECURITY AND THE GREAT POWERS

Barry Buzan
and
Gowher Rizvi
with
Rosemary Foot, Nancy Jetly,
B. A. Roberson and Anita Inder Singh

St. Martin's Press   New York

© Barry Buzan and Gowher Rizvi, 1986

All rights reserved. For information, write:
Scholarly & Reference Division,
St. Martin's Press, Inc., 175 Fifth Avenue, New York, NY 10010

First published in the United States of America in 1986

Printed in Hong Kong

ISBN 0–312–74714–4

Library of Congress Cataloging-in-Publication Data
Buzan, Barry.
South Asian insecurity and the great powers.
Includes index.
1. South Asia—National security—Addresses,
essays, lectures. I. Rizvi, Gowher. II. Foot,
Rosemary, 1948–   . III. Title.
UA830.B86 1986   355'.033054   86–1819
ISBN 0–312–74714–4

For
Fred Skinner
and
Margherita and Pietro Barolo

Contents

Notes on the Contributors	ix
Preface	xi

PART I   INTRODUCTION

1	A Framework for Regional Security Analysis *Barry Buzan*	3

PART II   THE DOMESTIC COMPONENT OF THE SECURITY PROBLEM

2	India: The Domestic Dimensions of Security *Nancy Jetly*	37
3	Pakistan: The Domestic Dimensions of Security *Gowher Rizvi*	60

PART III   THE REGIONAL COMPONENT OF THE SECURITY PROBLEM

4	The Rivalry Between India and Pakistan *Gowher Rizvi*	93
5	The Role of the Smaller States in the South Asian Complex *Gowher Rizvi*	127

PART IV   THE SUPER-REGIONAL COMPONENT OF THE SECURITY PROBLEM

6	South Asia and the Gulf Complex *B. A. Roberson*	159
7	The Sino-Soviet Complex and South Asia *Rosemary Foot*	181

PART V THE GLOBAL COMPONENT OF THE SECURITY PROBLEM

8 The Superpower Global Complex and South Asia 207
   Anita Inder Singh

PART VI CONCLUSIONS

9 The Future of the South Asian Security Complex 235
   Barry Buzan and Gowher Rizvi

   Index 253

Notes on the Contributors

Barry Buzan is a Senior Lecturer in International Studies at the University of Warwick. He obtained his PhD from London University for a study of the British peace movement between the wars. During the 1970s, he worked mostly on law of the sea issues, publishing widely on the process of negotiation, on Canadian policy, and on security aspects. Works from this period include *Seabed Politics* (1976), *A Sea of Troubles* (1978) and *Negotiating by Consensus* (1981). Dr Buzan now writes on the theoretical side of the national security problem in international relations and has published *People, States and Fear: The National Security Problem in International Relations* (1983).

Rosemary Foot is a Lecturer in International Relations at the University of Sussex. She obtained her PhD from the London School of Economics for a study of Great Power Rivalry in central West Asia in the 1960s. She has written extensively on the European community, Sino–Soviet rivalry and Sino–American conflict in Korea. She is the author of a forthcoming study, *The Wrong War: American Policy and the Dimensions of Korean Conflict 1950–1953*.

Nancy Jetly is a Lecturer in International Relations at the Centre of South, Southeast and Central Asian Studies at Jawaharlal Nehru University in New Delhi. She obtained her doctoral degree from Jawaharlal Nehru University and has been writing on South Asian affairs.

Gowher Rizvi lectures in International Studies at the University of Warwick. He obtained his DPhil from Oxford in Imperial and Commonwealth History. He has contributed several articles on themes in imperial policies, the problems of Muslim minorities, the institutionalisation of military regimes in Bangladesh and Pakistan.

He is the author of *Linlithgow and India: A Study of Constitutional and Political Impasse in India 1936–1943* (1978), and has co-edited *Perspectives on Imperialism and Decolonization* (1984) and *Indo–British Relations in Retrospect* (1984). He is currently working on a history of the British Empire (1776–1976) and has recently completed *Bangladesh: The Struggle for the Restoration of Democracy*. He has previously taught at the Universities of Oxford, Kent, and Canterbury in New Zealand.

B. A. Roberson is a Lecturer in International Relations at the University of St Andrews, and convenor of a Middle East Studies Group for the British International Studies Association. She is the author of 'The North–South Dialogue: Prospect for Development', *Millennium* (Special Issue, 1979), and is completing a doctoral thesis at London University on 'The Making of the Modern Egyptian State'.

Anita Inder Singh is a member of the research department of the Swedish Institute of International Affairs in Stockholm, and is working on Western interests in post-1947 India. She took her MA in History from the University of Delhi, and her MPhil from the Jawaharlal Nehru University. Her Oxford DPhil thesis, 'The Origins of the Partition of India, 1936–47', is due to be published in the Oxford South Asia series. She has taught History at the University of Delhi, and has been a Research Assistant with the Independent Commission on International Humanitarian Issues in Geneva. During 1982–4 she was a Visiting Research Fellow at the Graduate Institute of International Studies in Geneva.

Preface

The idea of this book originated when Gowher Rizvi, a student of Commonweath history with an interest in South Asia, encountered the newly evolved concept of 'security complexes' in Barry Buzan's *People, States and Fear* (1983). Security complexes seemed to offer a way out for a more meaningful analysis of regional security: it avoided the distortions of narrowly focused area studies and yet did not fall into the pitfalls of grandiose but vague generalities of security studies. The arguments for 'security complexes' as an analytical framework seemed intellectually compelling enough to be worth subjecting it to the test of an empirical study. This book, focusing on South Asia, attempts to do just that.

The framework of security complexes calls for analysis at three levels: national, regional and global. As neither of us had the expertise to undertake the entire task ourselves, we invited four other colleagues – Rosemary Foot, Nancy Jetly, B. A. Roberson and Anita Inder Singh – with their varied specialist knowledge, to join forces. Although this is a multi-authored work and each author approached the subject as he or she deemed best, we endeavoured to give the book rigour and focus by carefully defining the framework and posing specific questions to each contributor. In fact, we aimed to make the framework operational not only for this test study but also for anyone else engaged in the analysis of regional security.

We are grateful to the Centre for Indian Studies at Oxford and especially to its Director, Tapan Raychaudhuri, for organising a conference on 'India's Security: Past, Present and Future', where earlier drafts of many of the papers in this volume were first presented. We benefited immensely from discussions and comments and we should like to express our gratitude to all those who participated, but most particularly to Peter Lyon, Neville Maxwell, Gerry Segal and Krishnaswami Subrahmanyam.

As always one of us is deeply beholden to Sir Edgar Williams for his friendship and numerous kindnesses.

And finally, in more ways than we can recount, we are grateful to our wives: Deborah for lightening the strain of a preoccupied husband; and Agnese for her constant encouragement, sympathy and tolerance which made it possible to undertake this work amidst other commitments.

BARRY BUZAN
GOWHER RIZVI

Part I
Introduction

1 A Framework For Regional Security Analysis

BARRY BUZAN

One has only to read a newspaper, or cast an eye over recent publishers' lists in the field of international relations, to see that the demand for regional security analysis is high. The conflicts and crises that represent the most visible elements of international insecurity are nearly always described in regional terms. The crop of most current interest include Europe, the Middle East, the Gulf, Southwest Asia, the Horn, Southern Africa, Indo–China, and Central America.

At its crudest, this use of regional labels simply reflects a shorthand way of describing crises or conflicts involving more than one country. International security problems involve two or more countries by definition. And domestic security problems frequently spill over into the international domain, as illustrated most recently by the internal histories of Pakistan, Afghanistan and Iran.

But the idea of regional security is more important than mere situational name-tags, and more enduring than the passage of some particular war or confrontation. This deeper meaning expresses the sense that distinct and significant subsystems of security relations exist among some sets of states whose fate is that they have been locked into geographical proximity with each other. Such a sense has traditionally been very strong in thinking about the states of Europe. But because the European states came to dominate the world system – and indeed to occupy much of it directly – no similarly strong sense of other local security subsystems has developed within Western thinking about international relations. Even the massive process of decolonisation, which should logically have caused attention to be paid to the re-emergence of local security subsystems, has been overshadowed by the global rivalry between the superpowers. So weak is the sense of regional security patterns in Western thinking, that the authors of an otherwise excellent recent study on security in Southern Asia frankly admitted that their choice of a set of countries as a region was 'necessarily arbitrarily defined'.[1] The state of the literature on regions perhaps justifies this view, but the nature of the reality, in our opinion, does not.

This chapter contends that the lack of a stronger sense of regional security subsystems is a major weakness in contemporary analyses of international relations. In the absence of any developed sense of region, security analysis tends to polarise between the global system level on the one hand, and the national security level of individual states on the other. This tendency is exacerbated by the nature of expertise within the field, which also tends to divide between global system analysts and country specialists. Its consequence is that security analysis swings between an over-emphasis on the dominant role of the great powers within the global system, and an over-emphasis on the internal dynamics and perspectives of individual states.

Both of these extreme perspectives exaggerate the importance of the great powers in relation to local states. The global perspective does so because it focuses analysis onto a scale so large that only the great powers register as significant. The individual state perspective does so because it sets the relatively minor weight of single local states into direct contrast with the great powers. Both perspectives miss the regional level, which comprises the dynamic of security relations among the local states. This middle level is vital to the overall picture of security relations because it provides a strong mediating factor between the great powers and the local states. Unless that level is properly comprehended, neither the position of the local states in relation to each other, nor the character of relations between the great powers and local states, can be understood properly. The purpose of this chapter is therefore to elaborate a framework for regional security analysis that will define the sequence and orientation of the rest of the chapters in the book.

THE FRAMEWORK

Any attempt to study security has to face the problem of the seamless web. Security is a relational phenomenon. It involves not only the capabilities, desires and fears of individual states, but also the capabilities, desires and fears of the other states with which they interact. Because security is relational, we cannot understand the national security of any given state without understanding the international pattern of security interdependence in which it is embedded.

The trouble with the seamless web is that its logic pushes towards hopelessly complicated holistic perspectives. If the security of each is related to the security of all, then nothing can be fully understood without understanding everything. The reality of security interdependence is unavoidable. Consequently, the only hope of defining manageable subjects for study that neither lose, nor succumb to, the vital sense of the whole, is to find a hierarchy of levels within the holistic perspective. Each of these levels must identify durable, significant, and substantially self-contained features of the security problem. But no one of them will, by itself, be adequate to understand the problem as a whole, and the full meaning of each will only become clear in relation to the others.

The essence of our framework for regional security analysis is that it will specify, and attempt to define, a comprehensive set of such levels. By requiring the examination of all the levels, and the interactions among them, the framework should produce a balanced picture which avoids both the unmanageable complexities of the seamless web, and the dangerous distortions of perspectives weighted towards any one level. It should enable us to take any single country, and build up a picture of the successive layers from internal to global that define both its security context and its security problem. For the purposes of this book, the framework also enables the six contributors to combine their varied expertises in a systematic and complementary fashion. To the extent that this exercise is successful, we offer the framework as a tool for combining the intellectual coherence of single-author works, with the breadth and depth of expertise available in multi-author collections.

Two levels are readily apparent: the bottom one of the individual state, and the top one of the system as a whole. On the bottom level we are dealing with the basic unit of security in the international system. We can examine the security problem on this level both in terms of the inherent stability of the state itself, and in terms of its vulnerabilities to threats from outside.

On the top level we are dealing with the structure and characteristics of the entire international system. This is a much studied subject, and many different approaches to it exist that are relevant to the problem of security. Structural analyses of the international system can be found in both political and economic terms. Through these abstract perspectives it can be argued that the security problem is affected by changes in such variables as the distribution of power, the dominant mode of production, and the pattern of international trade.[2] The security character of the system can also be defined in terms of other highly generalised variables such as the nature of prevailing military technology, the degree of homogeneity or diversity in the dominant organising ideologies, and the strength of anti-war conditioning remaining in the system as a result of its last major bout of conflict.[3]

In this book, limits of space prevent us from dealing specifically with these sophisticated approaches to the security consequences of system structure. We will assume them to be generally understood as the background to contemporary international relations. We exclude them from the case study because they apply to the system generally, and are not in any way unique to South Asia. At this top level, we will focus simply on the dominant pattern of interactions among the great powers, and the way in which this pattern penetrates and influences the security interests of local states.

However, while the top and bottom levels are quite easy to define, the intermediate levels, which are vital to the balance of the framework, pose a problem. The principal candidate is the idea of region itself, or, put more technically, the notion of subsystems. If we could identify regional subsystems, then we would have two intermediate levels of analysis between system and state: the subsystems themselves, and the pattern of relations among them.

Unfortunately, the existing literature provides little guidance on how to define regional subsystems, although recent work by Vayrynen lays some potentially useful foundations.[4] By itself, the idea of subsystem – defined by Haas as 'any subset' of the international system[5] – is so broad as to be susceptible to almost any number of empirical interpretations. Haas and Brecher have both made attempts to devise frameworks for subordinate subsystems based on fairly broad notions of what constitutes a region. Similarly, Russett, and Cantori and Spiegel, have studied broadly defined regions from the perspective of integration.[6] But none of these approaches has resulted in any widely accepted definition of region. Part of the reason for their failure lies in the idealist, functionalist concerns of most of their authors. These concerns led to a search for the foundations for regional integration, which not surprisingly proved difficult to find. A review of the literature on regional subsystems by Thompson sadly catalogues the lack of progress towards development of any coherent theoretical or even descriptive framework.[7]

The only existing subsystem idea with any potential for the purposes of security analysis is the very traditional notion of local or regional balances of power. But this idea has never proved very useful precisely because it was confined to that single dimension – power – on which the great power dynamic most strongly overrode and obscured the local ones. Although local balances of power do operate, and are a significant feature of the security environment, they can be easily upset or distorted by movements in the globe-spanning resources of the great powers. Because of their susceptibility to external influences, balances of power are a much less reliable guide to structure in the periphery than they are at the centre.

Since we are interested in security analysis, a more fruitful way to approach the identification of regional subsystems is through the concept of security itself. Security is a broader idea than power, and it has the useful feature of incorporating much of the insight which derives from the analysis of power.[8] The principal dimension that we want to add to power is the pattern of amity and enmity. A hard view of balance of power theory would hold that patterns of amity/enmity are simply a product of the balance of power, with states shifting their alignments in accordance with the dictates of movements in the distribution of power. The view here is that the historical dynamic of amity and enmity is only partly related to the balance of power, and that where it is related, it is much stickier than the relatively fluid movement of the distribution of power. Consequently the two patterns must be considered as distinct factors in the security problem.

Specific patterns of amity/enmity may arise from local issues, like border disputes and ideological alignments, which could not be predicted from a simple consideration of the distribution of power. In pure power terms, a Sino–Indian alignment against the Soviet Union, and Soviet support for Pakistan, makes as much sense as the current pattern of alignment. By adding the dimension of amity/enmity to the picture, we get a clearer sense of the relational pattern and character of security than that provided by the pure abstraction of the balance of power view.

On this basis, we can attempt to define regional security subsystems in terms of patterns of amity and enmity that are substantially confined within some particular geographical area. The assumption is that local sets of states exist whose major security perceptions and concerns link together sufficiently closely that their national security problems cannot realistically be considered apart from one another. In order to differentiate this type of set from other ideas of subsystem, I propose to call such a set a *security complex*. This name has the advantage of indicating both the character of the attribute that defines the set (security), and the notion of intense interdependence that distinguishes any particular set from its neighbours. Unlike most other attempts to define regional subsystems, security complexes rest, for the most part, on the interdependence of rivalry rather than on the interdependence of shared interests.

A security complex represents more than just the arbitrary abstraction implied in the 'any subset' definition of subsystem. Security complexes are an empirical phenomenon with historical and geopolitical roots. They are characteristic products of an anarchic international system, and represent durable rather than permanent substructures within such a system. Although this argument courts the charge of reification (treating abstractions as objects), the fact that security complexes are in some sense 'real' is important, because it is their reality which explains the mediating effect that they have on relations between the great powers and the local states.

The reality of security complexes lies more in their parts (the individual lines of amity and enmity between states) than in their wholes (the notion of a discrete subsystem), and this feature complicates the process of identifying them in hard scientific terms. The lines of security concern can be traced quite easily by observing states' foreign policy and military behaviour: an obvious example being the way Pakistan deploys the bulk of its army against India rather than on its borders with China, Afghanistan or Iran. But assessing the overall pattern formed by those lines, and particularly finding concentrations of interaction within the pattern which are strong enough to constitute a complex, may be a matter of controversy. The problem of the seamless web is at its most acute in the middle levels of analysis.

There are bound to be empirical disagreements about the strength of given lines of security interdependence. Measuring variables like amity and enmity, is an even less precise business than measuring power, though in both cases the main features usually stand out quite clearly even in the absence of scientific scales for comparison. Such disagreements are unavoidable, and they may well lead to differing interpretations about how a given complex should be defined. For example, should the Gulf and the Middle East be considered as two complexes or as one? To some extent each area has an identifiable security dynamic of its own, but their dynamics also overlap in important ways, especially because of the linking position of Iraq. Controversy of this type reflects genuine ambiguities in the real world. What is important for the framework overall is that *some* coherent sense of the regional security dynamics be interposed between the global and state levels. The main issue is recognising that strong local security dynamics exist. Once that task is undertaken as part of the framework, disagreements about the location of boundaries within the seamless web are unlikely to result in major contradictions about the main lines of amity and enmity. Indeed, by providing alternative perspectives, they could well enrich analysis where genuine ambiguity exists.

The task of identifying a security complex involves making judgements about the relative strengths of security interdependencies among different countries.[9] In some places these will be very strong, as between India and Pakistan, in others relatively weak, as between Bangladesh and Burma. Within the seamless web of security interdependence, we can thus expect to find patterns arising from the existence of stronger and weaker lines of amity and enmity. A security complex exists where a set of security relationships stands out from the general background by virtue of its relatively inward-looking and self-contained character. Security interdependencies will tend to be more strongly focused among the members of the set than they are between the members and outside states. The boundaries between such sets will thus be defined by the *relative indifference* attending the security perceptions across them. The strong security links between India and Pakistan put them clearly within the same complex, while the relatively weak links between Iran and Pakistan suggest the existence of a boundary between complexes.

The principal factor defining a complex is usually a high level of threat which is felt *mutually* among two or more major states. Unless they are world class powers, these states will usually be close neighbours. In theory, a high level of trust and friendship can also serve as a defining factor, as it does in the ASEAN component of the Southeast Asian complex, but instances of this are rare. In South Asia, the complex is defined by the Indo–Pakistan rivalry, while in the Gulf, it is defined by a three-sided rivalry among Iran, Iraq, and Saudi Arabia. The extent to which neighbouring local dynamics like these two are distinct is indicated by the rhetoric of the states concerned, by their military deployments, and by the record of their conflicts. It cannot be without significance in terms of defining the pattern of local security dynamics, that the various wars within each of these two sets have had little military or political impact on the members of the other set, and that neither their behaviour nor their rhetoric indicates any significant perception of threat across the boundary between them. Similarly, there has been little interaction between the wars in South Asia, and those in Southeast Asia.

Because security complexes are geographical entities, they will often include, by default, a number of minor states. Due to their relatively low power in comparison with their neighbours, these states may have little impact on the structure of the complex. Their own securities are intimately bound up in the pattern of the larger states, but they can only become a source of threat to a larger state by virtue of the impact of their alignments on relations among the larger powers. In the Gulf, Kuwait, Bahrein, Qatar and the UAE, are in this position. In South Asia, it is occupied by Bangladesh, Bhutan, Nepal, Sri Lanka, and the Maldives, nearly all of which are tied closely to India by the dictates of geography and culture. A few weak countries may also occupy buffer positions between neighbouring security complexes. Their influence on affairs in either complex will be relatively marginal, and they exist in those areas where the larger patterns of security orientation stand, as it were, back to back. Burma is a very clear example of this role between South and Southeast Asia, as is Turkey between Europe and the Middle East. Afghanistan has traditionally been seen as a buffer, but its classification in this way is more arguable, both because of its substantial disputes with Pakistan, and recently because of its occupation by the Soviet Union.[10]

The question of the role of cultural and racial ties in the definition of security complexes is an interesting one. It seems not unlikely that shared cultural and racial ties among a group of states would cause them both to pay more attention to each other in general, and to legitimise mutual interventions in each other's security affairs in particular. The South Asian complex coincides with what K. Subrahmanyam has called a 'civilizational area'.[11] It is not difficult to find evidence that Arab and Islamic cultural factors likewise facilitate and legitimise security interdependence among a large group of states. And a similar line of thinking underlies much traditional analysis of European history, with its emphasis on the role of Christendom in defining a community of states. In analysing the shape and structure of security complexes cultural and racial patterns may well be an important contributing factor. But they come second to the patterns of security perception which are the principal defining features of a security complex.

One further difficulty in the identification of a security complex is the existence of a lopsided security link between two major local states. The relationship between China and India provides a clear example of this problem. China is a major security concern for India: arguably even the principal one. Chinese military strength lies close to India's main centre of population, and China holds territory claimed by India. But by itself, India is a relatively minor security concern for China. Indian military strength sits far from China's heartlands, and weighs little compared with other threats to China's interests.

A situation like this typically indicates the existence of a boundary between a *lower* and a *higher* level security complex. A lower level complex is composed of local states whose power does not extend much, if at all, beyond the range of their immediate neighbours. This constraint on power is a key element in the existence of relatively self-contained local security dynamics among sets of neighbouring states. A higher level complex, by contrast, contains great powers: states like the United States and the Soviet Union, whose power may well extend far beyond their immediate environment, or states like China and the Soviet Union, whose power is sufficient to impinge on several sectors of what their enormous physical size makes a vast 'local environment'.

This distinction between lower and higher level complexes enables us to sketch out the full range of levels that comprise our framework. At the bottom end we have the internal security environment of individual states. Next comes the lower level, or local, security complexes. We would expect security relations to be relatively intense within these complexes, and relatively subdued between them. But in some instances, as we shall see in more detail later, we will find a number of local complexes that are linked together in a complex of complexes. This is true, for example of the set of local complexes which incorporate the more than twenty Arab states, as well as Israel, Ethiopia and Iran. At least four distinct complexes occur in this area: the Gulf, the Middle East, the Horn, and the Maghreb. But all of them are linked together within the broader framework of Arab and Islamic politics. That framework itself comprises a significant security boundary between the states within it and those outside. Wherever we encounter such an aggregation of complexes, we will refer to it as a *supercomplex*. At the top end we find the higher, or great power, complexes. These define the system level, as discussed above. We would expect security relations among the great powers to be intense, and to penetrate in varying degrees into the affairs of the local complexes.

In principle, this pattern could cover the entire membership of the international system. In practice, however, some areas may exist where the local states are so weak that their power does not project much, if at all, beyond their own boundaries. In relation to each other, these states have inward-directed security perspectives, and there is not much interaction between them. Under these conditions, which are characteristic of some parts of Africa and also among the small Pacific Ocean island states, a local security complex can only be said to exist in a very weak and shadowy form.

Like a balance of power, a security complex can exist and function regardless of whether or not the actors involved recognise it. They will, of course, recognise the particular lines of amity, enmity and threat which bear on them, for it they did not, the whole idea of security complexes would be void. But they may well not see, or appreciate fully, the whole pattern of which they are a part. Typically, they will be much more aware of the threats others pose to them than they will be of the way they are seen by others. This issue of recognition marks a fairly sharp analytical divide between balance of power models and subsystems, with the former generally discounting it, and the latter frequently requiring it as part of the definition of a subsystem.[12] Though recognition of the complex, as with a balance of power, is not a necessary condition for its existence, if recognition occurs, it may well influence the policies of the actors involved by making them more conscious of the structural context underlying their specific policy problems.

In looking for the sets of states that constitute security complexes, we are primarily concerned with the military and political dimensions of security. The reasoning behind this selection is that it is these dimensions that are the most relevant to the patterns of threat and amity/enmity that define the set. For any given state; the problem of economic security is likely to have quite a different relational dynamic from that of its military and political security. In the European complex before 1914, for example, major military rivalries ran alongside extensive trading relationships. But in the contemporary superpower complex comprising the United States and the Soviet

Union, a strong interdependence in military security is paralleled by a very low level of economic exchange. A pattern similar to that of the superpowers prevails between India and Pakistan in South Asia. Under such conditions, the economic security of the states concerned does not depend primarily on their relationship with the other states within the complex. This disparity between economic security on the one hand, and political and military security on the other, occurs frequently. The reason for it is that economic relations are not nearly so much conditioned by geographical proximity as are political and military ones. This geographical factor explains why lower level security complexes nearly always comprise territorially coherent areas.

Economic factors do play a role in determining both the power of states within their local complex, and their domestic stability and cohesion as actors. They may also play an important role in motivating the patterns of external interest in the local complex, as in the case of the United States and the oil-producing countries of the Gulf. And they can affect the prospects for regional integration, which, as we shall see, can influence how a given security complex evolves. South Asia, for example, offers much less prospect for economic cooperation than does Western Europe, because the similarities among the South Asian economies make them more competitive than complementary. So, economic factors need to be taken into account in analysing a security complex, but the framework we are using here is not appropriate for examining economic security itself.[13]

THE FRAMEWORK APPLIED TO SOUTH ASIA: AN OVERVIEW

The method of analysis within the framework is first to understand the security dynamic at each level, and then to see how the patterns at each level interact with each other. This book is structured around a hierarchy of four levels: domestic, regional, super-regional, and global.

The domestic level concerns the two main actors in South Asia, India and Pakistan. How strong and stable are they as political units, and how do their domestic characteristics relate to their external security problems?

It is vital to understand the domestic character of the states concerned, because states with weak and unstable internal structures clearly have more vulnerabilities, and therefore face different and more complicated security problems, than those with a high degree of internal unity and consensus. Politically weak states will be particularly vulnerable to, and therefore particularly sensitive to, political threats, regardless of their military strength. States as political units are very varied in character. Some, like Chad, Lebanon, and Laos, have such weak political structures that they barely qualify as states at all, except that they enjoy the recognition of most other states as a legitimate member of the class. Others, like France and Japan, enjoy the benefits of a strong and deeply rooted synthesis between nation and state. On this scale, Pakistan is substantially more towards the weak end than is India. We need to understand the reasons for this differences, and to relate them to the dynamic of the South Asian complex as a whole.

The regional level concerns the relations among the states within a local security complex, in this case, the South Asian one. Local complexes differ in terms of their size, structure, and intensity and stability of internal relational pattern. The South Asian one is probably average in size, but because it is centred on only two local powers, it is relatively simple in structure. Relations within it have been quite intense, and partly for that reason, its structure has endured for several decades, and it is relatively easy to define. It came into being with the British withdrawal from the region in 1947, and right from that start its principal defining feature was the hostility between India and Pakistan. These two countries were born locked into a complicated rivalry that defined the central security problem for each of them. They easily overawed the smaller states which were geographically entangled within their sphere, and so fell naturally into a power rivalry with each other. Afghanistan and Burma provided buffer zones at either side of the complex, insulating it from the neighbouring local security complexes in the Gulf and Southeast Asia. Five of the states within the complex shared the long border with China, but a buffer of sorts was provided here both by the Himalaya Mountains, and by the fact that the great weight of the Chinese population was located and oriented much further to the east.

The rivalry between India and Pakistan stemmed not just from Pakistan's desire to avoid being overawed by India, but also from several specific disputes arising from their former integration. A major territorial dispute in Kashmir, and a lesser one in the Rann of Kutch remain as the heritage of a hurried partition. These disputes are not only territorial, but reflect an enduring political tension resulting from the principle of partition. Pakistan was created as an Islamic state, and commonality of religion provides the major binding force among its diverse national groups. India is organised on the principle of secular federalism, although Hindus are by far the largest group. It still contains a very substantial population of Muslims, some of whom inhabit the areas disputed with Pakistan. Thus, the organising principle of Pakistan threatens India with secessionism, while that of India threatens Pakistan with either dismemberment or absorption. These structural threats interact importantly with the domestic political weaknesses in each country.

Our task at the regional level is thus to understand the security dynamic of the local complex in its own right, and to relate it both up and down to the dynamics of the other three levels. To do this we will examine both the rivalry between India and Pakistan, and the role of the smaller states.

The super-regional level consists of the relations among adjacent security complexes. The South Asian complex borders on three others, two, like itself, lower level, local complexes, and one higher level, great power complex. To the west, lies the Gulf complex, and the larger Middle Eastern supercomplex of which it is a part. To the east lies the Southeast Asian complex. And to the north lies the Sino-Soviet complex. Although there is a local dynamic of security which is largely self-contained within each complex, there is often some interaction across the boundaries, especially between higher level and lower level complexes. This interaction needs to be understood, especially if it is of a type or extent to call into question the validity of the boundary. As we shall see in Part IV, the possibility of merger between all or part of two complexes constitutes a marker of significant change.

Of the two lower level complexes that border on South Asia, the Gulf/Middle Eastern is much more important than that in South-east Asia. South Asia has remained almost completely insulated from the turbulent affairs of South-east Asia, and although the two complexes share Burma as a buffer, their impact on each other is negligible. The only connection of any consequence between them has been India's diplomatic support for Vietnam, particularly its controversial recognition in 1980 of the Vietnamese-imposed Heng Samrin government in Kampuchea. India's action reflects both its alliance with the Soviet Union, which is also Vietnam's major ally, and its fear of China, where India's interest is served by the distraction of China's attention towards rivalry with Vietnam.[14] This condition of relative indifference between the security dynamics of South and Southeast Asia appears stable. Change would require the development of a major rivalry between India and China, or India and one of the Southeast Asian powers, neither of which seems likely for some decades to come. Because of the small scale of interaction between the South and Southeast Asian security complexes, the latter will not receive any detailed or separate treatment in this study.

The Middle Eastern supercomplex, by contrast, is in some ways quite closely attached to South Asia, not least by the links of Islam. The Middle Eastern supercomplex, as indicated above, is much larger, and has a much more complicated structure, than its South Asian neighbour. Although it is the Gulf complex which has the most bearing on South Asian affairs, we cannot ignore the broader context of the supercomplex. The Gulf countries are active participants in the affairs of other Middle Eastern complexes, just as the other Arab states also play a role in the affairs of the Gulf. In addition, the reactivated and overarching embrace of Islam creates an area of significant political and military linkage between Pakistan and the supercomplex as a whole.[15]

Until quite recently, the security dynamics of these two local complexes made little impact on each other. The numerous wars within each had no particular effect on relations within the other, and despite some interaction between the two, there was no reason to question the essential integrity of their separate security dynamics. For a time during the 1970s it appeared that the Shah of Iran's expansive regional ambitions might begin to clash with India's sense of itself as the major regional power. This possibility for a shift in security orientations, however, seems largely to have disappeared with the fall of the Shah in 1979.

Perhaps more important in the long run is Pakistan's increasing cultivation of Islamic links, particularly with the Gulf Arab states. This trend has been active since the early 1970s, after the breakaway of Bangladesh, as part of Pakistan's attempt to maintain the balance against India.[16] The question is whether these developments are strong enough to call into question the boundary between the South Asian and Gulf complexes. Do we need to redefine the basic local structure of security relations within the area of South and Southwest Asia?

An additional factor bringing pressure to bear on the separateness of the two complexes is the Soviet occupation of Afghanistan since

1979. This event transformed what had been a remote buffer into a focal point of major tension between the superpowers. As a centre of international rivalry, Afghanistan forces connections between the two local complexes as much as it serves to separate them.[17] The direct intrusion of Soviet power into the region raises important questions about the interaction between the local and the higher level security dynamics. From the geostrategic perspective of the superpowers, the relevant region incorporates both the sub-continent and the Gulf. This perspective clashes with a local structure which is still based on two distinct regional complexes. The question is whether the superpower impact will be sufficient to create, or hasten, a change in the local structure, or whether the local structure will endure, keeping Afghanistan largely contained in its traditional role of buffer?

The Soviet occupation of Afghanistan underlines the fact that South Asia shares a long border with a higher level complex, the Sino–Soviet. Such a situation is not unique to South Asia, but it does create special circumstances that are not common to all local complexes. Partly because of the geographical proximity, we have decided to discuss the Sino–Soviet complex on the super-regional, rather than on the global level. A case could be made for including China on the global level because of its key position in the Soviet–American balance. But China's power is not yet, in itself, global in reach, and its rivalry with the Soviet Union is beginning to take on form of an Asian supercomplex. The odd position of China means that some confusion will result no matter which level it is included within. On balance, our assessment is that the global level is not yet tripolar, and that the super-regional level is therefore the most appropriate.

The Sino–Soviet security complex came into play during the early 1960s, as a result of China breaking away from the alliance with the Soviet Union that had locked it into the Soviet–American superpower complex during the previous decade. The rivalry between the two communist giants spilled over into South Asia almost immediately, the pace and scale of the intervention being much increased by the nearly simultaneous peaking of a quite independent border crisis between India and China. For the Soviet Union, the split with China simply added reasons for continuing its efforts to cultivate closer relations with India. India's fear of China after 1962 suited Soviet purposes well, since a strong, Soviet-armed India would serve as a bastion in Soviet attempts to contain China's influence.

China's obvious response to the growth of Indo–Soviet friendship was to strengthen relations with Pakistan, and an informal alliance was in place between the two by 1963. The Chinese became Pakistan's principal supplier of arms, even though their limited industrial capacity prevented China from matching the technological sophistication of the superpowers. Perhaps more importantly, the Chinese provided political support for Pakistan in its wars with India, giving the Indians the constant worry of a two-front war.

The Sino-Soviet penetration into South Asia is part of a larger pattern in which their rivalry is becoming an increasingly prominent force in all three of the peripheral Asian security complexes (South Asia, South-east Asia and North-east Asia). The rise of Chinese power, and its shift from alignment with the Soviet Union to opposition, radically changed the superpower game in Asia by creating a powerful, independent, and locally based centre of opposition to the Soviet Union. Although American power is still firmly established in the rimlands of Asia, it has withdrawn offshore in South-east Asia, and is only tenuously established in Pakistan.

On the assumption that China's relative power continues to increase, and that the Sino–Soviet rivalry remains in being, we could be witnessing the emergence of an Asian supercomplex. This supercomplex would be defined by a pattern of Sino–Soviet rivalry in all three peripheral Asian complexes. The Soviet Union would be trying to contain and counter-balance China, and China would be trying to assert its own sphere of influence, and to prevent Soviet penetration of the countries on its borders. This scenario depends only on China and the Soviet Union continuing to see themselves as rivals. As with the United States and the Soviet Union, if that condition is met, whether the rivalry is characterised by *détente* and peaceful coexistence, or by confrontation and crisis, makes little difference to the incentives on both sides to compete for allies and spheres of influence. Japan, should it again decide to take an independent role in international affairs, could either reinforce China's position or else find its own distinctive role as a local great power.

The global level consists of the higher level security complex which centres on the rivalry between the United States and the Soviet Union. These two superpowers define the top level of the global power structure. Although their primary security concerns are with each other, the range of their power means that they project their rivalry into all the other complexes in the international system. This projection may be relatively mild, as it was in most of Africa during the 1960s, or it may be overwhelmingly intense, as in Europe since 1945.

The full arrival of the Soviet–American security complex onto the world stage is conventionally placed in the same year, 1947, as that of the South Asian. The global competition of the two superpowers needs no general description here. Suffice it to say that their political and military rivalry quickly took on global dimensions which resulted in extensive interventions in the affairs of local complexes everywhere. The American policy of containment defined a huge arc of confrontation, stretching from Western Europe, through the Middle East, South Asia, and Southeast Asia, to Korea and Japan in the Far East. Along this arc, the imposition of the global complex on the local ones was much more intense than it was in Africa and Latin America. In Europe and Northeast Asia, the superpowers effectively overlaid and suppressed the local security dynamics by a policy of collective alliance and direct military presence. In the three southern complexes, superpower competition was played primarily in terms of alignments and alliances, although the United States tried direct presence in Southeast Asia, and the Soviet Union is currently trying it in Afghanistan.

The impact of the global complex on the South Asian one goes back to the early 1950s, and has been both continuous and important. In effect, the superpowers have added the momentum of their own rivalry to the already existing rivalry between India and Pakistan. After much beckoning from Pakistan, the United States finally drew it into its network of containment alliances in the early 1950s. During the 1960s, Pakistan faced the difficulty of having its two principal supporters at loggerheads. But this problem was solved in the early 1970s, with active Pakistani assistance, by the Sino–American *rapprochement*. This *rapprochement* effected a substantial merger of the Sino–Soviet and Soviet–American complexes within South Asia. Although the global dynamics of these two higher level complexes remained fairly distinct, in South Asia it became reasonable to talk about a Sino–Pakistani–American axis in general opposition to an Indo–Soviet one.

The Soviet penetration into South Asia came somewhat later than the America one. Because India was, during the 1950s, both complacent about its military strength and wedded to a policy of nonalignment, the scope for a countervailing Soviet role in South Asia was limited. The Soviets, like the Americans, saw their best opportunity for penetration in terms of arms supply. By offering modern arms at relatively low cost, the Soviets were able to establish themselves as India's principal supplier from 1962 onwards. The shock to India of its defeat by China in 1962 did much to open the way for the Soviets because of the concern which it generated in India to upgrade and expand its military forces. Arms supply, and support for India against both the United States and China, resulted in the Indo–Soviet alliance of 1971.

This alliance did not draw India closely into the Soviet bloc, and India managed to retain, although with increasing difficulty, its status as a non-aligned country. But it did establish a firm pattern in which the alignments in the local and global complexes complemented and reinforced each other. The most important impact of this penetration was the injection of substantial supplies of modern armaments into South Asia. These armaments expanded the threats that India and Pakistan posed to each other, and played an important role in the 1965 and 1971 wars. It cannot be said that the dynamics of the South Asian complex have made much return impact on the affairs of the superpowers.

From the Soviet point of view, cultivation of Delhi served both to counter the Chinese, and to drive a wedge into the American containment system. The importance to the Soviet Union of a good relationship with New Delhi should not be underestimated. Its value to Moscow is indicated by the extent of the resources and commitments that the Soviets have devoted to keeping India at least neutral, and at best mildly pro-Soviet, in orientation. The Soviet's task has been much facilitated by the Sino–Indian dispute, and by the unwillingness of the United States to match the degree of respect that the Soviet Union has shown towards India's desire to be accorded the status of a major power.[18] Without the Indian link, the Soviet Union faces the prospect of virtually complete encirclement by major hostile powers, an encirclement that would exist even if the United States was not encouraging it.

From this brief survey, we can begin to see the utility of the layered framework as an analytical approach to the problem of regional security. We can take in all four of the relevant levels – domestic, regional, super-regional and global – while both maintaining the distinctions among the dynamics which define each level, and identifying the interactions between the levels. The broad patterns which have so far defined the context of security for South Asia emerge clearly. On the local level, there is a bipolar rivalry which has proved durable both because of the zero-sum nature of the hostility, and because of its linkage to domestic instabilities within both India and Pakistan. This rivalry is reinforced both by a link between Pakistan and the Gulf Arab states, and by two overlapping sets of penetrations from higher complexes, one of which has the added emphasis of a dispute across the boundaries of the complexes concerned.

Within the framework, the idea of security complexes offers some firm middle ground between the state and system levels of analysis. Although its application is limited to the confines of security issues, it provides us with some relatively concrete guidelines for defining the unit of analysis at the regional level. The concept thus also enables us to define the pattern of interaction among the subsystems – both laterally (i.e. between lower level complexes) and hierarchically (i.e. between lower and higher level complexes) – which complete the picture of the entire international security system. It is important that security complexes be seen as real units within the structure of the international system. They are obviously a looser class of entities than states, but the justification for seeing them as empirical, as well as abstract, constructs is that they have visible structures which are historically durable. If we accept them as meaningful units, then it becomes possible to use their evolution to identify significant change in the international system.

The next section concentrates on the security complex as an empirical unit of analysis. It examines both the mechanisms by which security complexes change, and the outcomes that such change can produce. Our purpose here is partly theoretical – to develop a dynamic dimension to what would otherwise be a static framework – and partly to establish the categories that will be used to organise the discussion in the final chapter of the book. In that chapter we want to speculate about the future of security in South Asia, and it is only by establishing a framework for change in the structure of security complexes that we can do so systematically.

SECURITY COMPLEXES AND THE DYNAMICS OF CHANGE

As with any discussion of change, we face the core theoretical problem of distinguishing what is significant from what is not. Change is ubiquitous and continuous, and yet we can say with meaning that on some levels some things remain effectively the same over long periods.[19] The South Asian security complex, for example, has displayed a durable structure for nearly four decades. Its continuity resides in the primacy of India and Pakistan within the subcontinent, in the continuation of their treatment of each other as major rivals, and in the absence of any developments within, or external impositions upon, South Asia strong enough to break this pattern. Structure in this sense has endured despite the endless changes within and around the subcontinent. Internal changes, including the partition of Pakistan, have not altered the basic pattern of relations. Neither have external ones, which have so far tended more to reinforce, than to undermine, the existing bipolarity.

We can use the notion of *essential structure* as a standard by which to measure significant change in a loose and decentralised entity like a security complex. The two key components of essential structure in a security complex are: (1) the patterns of amity and enmity, and (2) the distribution of power among the principal states. Major shifts in either would normally require a redefinition of the complex, although, as we shall see, the impact of power versus hostility shifts will vary according to the nature of the complex concerned.

The Sources of Change

With this idea of essential structure in mind, we can identify the sources of change most likely to lead to the redefining of a security complex either because of a shift in the distribution of power, or because of a shift in the pattern of amity and enmity. These changes can result from either internal or external factors, and more than one may be in operation at any given time.

Shifts in the Distribution of Power

Power shifts resulting from internal factors can be caused in several ways. Actors can disintegrate, like Pakistan in 1971, and Austria–Hungary after the First World War, or they can merge, like the component parts of Germany and Italy during the nineteenth century. But such events have to be on a large scale to affect the essential structure of the complex. In South Asia, India's absorption of small states like Goa and Sikkim, made no real difference to the distribution of power within the sub-continent. Pakistan's loss of Bangladesh was both absolutely and relatively a much more substantial event, and at first seemed likely to put India into a position of such dominance as to call into question the basic bipolarity of the distribution of power. But although Pakistan was certainly weakened politically by the loss of its pretention to Islamic exclusivity, its military strength was not much affected, and its leaders quickly left behind any inclination they might have had to defer to Indian hegemony. Despite its many troubles, Pakistan has remained both strong enough and wilful enough to sustain bipolarity. Only if the secession of Bangladesh turns out to be the first stage in a more complex balkanisation of Pakistan will the 1971 events be seen as the start of a major transformation of the South Asian complex.

Power shifts can also result from differences in the rate of development among actors. Japan rose to prominence in Asia because of its ability to modernise faster than its neighbours, and the Shah of Iran clearly had similar hopes for his own country. In the absence of different rates of growth, there is some scope for power shifts in the way countries allocate their existing resources. Major diversions of resources away from investment and civil consumption towards military expenditure can produce a power shift as India demonstrated in the doubling of its military budget after the 1962 war with China. But neither differences in rates of development nor differences in resource allocation have so far eroded the essential structure in South Asia. Rates of growth have not been wildly disparate, and while India has not tried to push its advantage of size, Pakistan has shown willing to bear a proportionately greater military burden.[20]

Power shifts resulting from external factors are similarly diverse. External actors can change the power structure of a local complex in two ways: either by joining it, if they are adjacent, or by making alignments within it, whether they are either adjacent and/or members of a higher level complex. For new actors to join a complex normally requires a redefinition of more than one complex, with the exception that small states may not make much impact on the essential structure. The historical and geographical ties which bind a local complex together are not easily overcome. In South Asia, one can imagine scenarios in which India achieves ascendency over Pakistan, and extends its influence into the Gulf, or in which some of the Gulf states develop such close ties to Pakistan that the security dynamics of the Gulf and South Asia become entangled into an expanded new complex. But neither of these possibilities has so far developed to the point of overturning the established patterns in South and South-west Asia.

External actors can impinge upon the local distribution of power in many ways short of actually joining, and thereby redefining, the complex. Since there are only two major powers in South Asia, the local context for external penetration is much simpler than, say, in the Middle East. In South Asia, external influence on the distribution of power has virtually all been concerned with using alignment and arms supply to raise the power of either India or Pakistan. The relatively even-handed influence of the ex-colonial power, Britain, waned fairly quickly, to be superseded by the more partisan and competitive interests of the United States, the Soviet Union, and China. This pattern of external penetrations into South Asia has not been aimed at overthrowing the essential structure of the complex, and has not been so one-sided as to upset the local balance. As the weaker local state, Pakistan has benefited more conspicuously than India, and it is worth asking whether Pakistan could have sustained its role within the bipolar structure of the local complex without the measure of external support it has received.

It becomes clear that external actors can impinge in several different ways. They can add to the power of local states directly, either by arms supply, or by the direct involvement of their own armed forces. Through economic aid, they can try to alter the local rates of development to the advantage of one side or the other. In the early stages of external penetration the addition of new supporters to one side or the other is likely to be a major source of influence on the affairs of the local complex. Later, the local balance may be affected by the withdrawal of one external actor, like the loosening of American support for Pakistan in the 1960s and 1970s.

In extremis, external powers can overlay the local power dynamic completely by imposing their own direct presence on the entire complex. Something like this happened in both Europe and Northeast Asia after the Second World War. South Asia experienced it up to 1947 in the form of British imperialism. The Raj suppressed the local security dynamic, and incorporated the sub-continent directly into the global pattern of great power rivalry, particularly Britain's 'containment' of southward Russian expansion. After independence, South Asia, unlike the Middle East, was notable for the absence of direct presence by outside powers. External inputs to the local distribution of power, with the exception of China, consisted almost entirely of indirect military and economic aid. China was an exception because of its direct dispute with India. Although Chinese forces have not been stationed in Pakistan, their presence on India's border has divided India's military attention, and thereby provided a useful support for Pakistan. Since 1979, Soviet troops have been in direct occupation of Afghanistan, but it is still too early to tell whether the overlaying of this peripheral member of the complex by an external power will transform relations in the sub-continent.

Shifts in the Pattern of Hostility

Assuming that the membership of a security complex remains stable, internal sources of shifts in the pattern of hostility occur either because an existing dispute has been resolved, or because new disputes have developed. Since there is only room for one major dispute within South Asia as it is currently constituted, an internal change in the pattern of hostility can only come from some form of resolution between India and Pakistan. Either the two countries can weary of their rivalry, and negotiate an agreement, as has been done by France and Germany, and Egypt and Israel, or else one side, most probably India, can effectively win the rivalry by overawing the other in some way. Alternatively, if the membership of the complex changes because of the break-up or merger of existing members, then a new pattern of amity and enmity occurs by definition. Some possibility for this exists in South Asia because of the internal fragility of Pakistan. Should Pakistan disintegrate into several smaller states, then India would unquestionably dominate the subcontinent. In that event, the structure of the complex would shift from bipolar to unipolar.

In complexes with three or more centres of power, like some of those with the Middle Eastern supercomplex, there is more scope for change in the pattern of hostility than there is in South Asia. Significant bilateral conflict resolutions, like that between Egypt and Israel, can occur without altering the overall pattern, in this case Arab–Israeli enmity, beyond recognition. Similarly, new conflicts can emerge to alter the priorities given to old ones. Iraq, for example gyrates among Syria, Iran, Saudi Arabia and Israel as its principal foe.

As a rule, external actors have a much lesser impact on the pattern of local hostilities than they do on the distribution of power. As the case of South Asia since 1947 illustrates, external actors tend, whether explicitly or implicitly, to fall into line with the pattern of local hostility. Since external actors are usually pursuing their own interests, acquiescence in the local pattern of hostility is much the easiest way to penetrate a local complex. Unless the pattern of intervention is very lopsided, competing external powers will therefore normally reinforce rather than change the existing pattern of local hostility. They may attempt to mediate in the local disputes, as the Soviet Union has done in South Asia, and the United States is still trying to do in the Middle East. But attempts by external powes to go against the grain of local alignments do not have a good record, and seem always clearly subordinate to continued support for local allies.

The major exception to this rule is when external powers resort to direct overlay of the local complex. The evidence we have indicates that external actors can only reliably change local patterns of hostility when they impose their own presence on the countries concerned. The American role in the peace between Israel and Egypt provides a minor example. Other than the colonial period, the principal illustration is the complete overlaying of the European security complex by the larger pattern of hostility between the United States and the Soviet Union after the Second World War. The direct presence of the superpowers has effectively suppressed virtually all of the previously existing local patterns of hostility for more than three decades. Only the hostility between Greece and Turkey has managed to prevail openly against it. The process of overlay, however, requires exceptional historical circumstances.

The generally low impact of external actors on local hostilities is indicated by the fact that the superpowers can even switch sides without affecting them as happened between Ethiopia and Somalia during the late 1970s. The only other way in which external actors can change the local pattern of hostility is by joining the complex. The logic of this is the same as that for power discussed above. Although it might be possible for external actors to affect local hostilities by manipulating changes of government in local actors, this course is uncertain in the extreme. Ethiopia and Iran are only the most recent examples of cases where even revolutionary changes in government do not result in alterations of the basic patterns of local security perceptions, though as the Iranian case illustrates, the rhetoric and the detail of those perceptions may change substantially. Indeed, it is a notable feature of contemporary superpower interventions in the domestic politics of Third World countries that the superpowers are much more interested in manipulating Third World attitudes towards the East–West divide, than they are in reconditioning the attitudes of Third World states towards each other.

The Outcomes of Change

The internal and external changes bearing on any given local security complex are usually both numerous and continuous. The key question is do they work to sustain the essential structure or do they push it towards some kind of transformation? Four broad structural options for assessing the impact of change on a security complex are available: maintenance of the status quo, internal transformation, external transformation, and overlay. Here, our task is simply to define these options. In Chapter 9 we will apply them to an assessment of South Asian security.

Maintenance of the Status Quo

Maintenance of the status quo means that the essential structure of the local complex – its distribution of power, and pattern of hostility – remains fundamentally intact. For this outcome to occur does not mean that no change has taken place. Rather, it means that the changes which have occurred have tended, in aggregate, either to support, or else not seriously to undermine, the local structure.

South Asia is a good example of a local complex in which the status quo has been maintained. The complex was born abruptly when the unifying overlay of British rule was replaced by a set of independent states. In this classic example of a system change from unitary to anarchic structure, war followed almost immediately, and since then, the continuity of the complex's essential structure has been demonstrated by a series of wars. None of these has been decisive or destructive enough either to break the power of one side, or else to convince both that their mutual hostility is self-defeating. Consequently, the South Asian wars have served more to reinforce existing hostility than either to eliminate it, or to break the bipolar power structure.

Several powerful external actors have brought their influence to bear on the complex, but again, the sum of their impacts has been to sustain the existing structure. It might be argued that this external support has become a vital prop to the status quo. If the subcontinent had been left to its own devices, the greater weight of India and the greater fragility of Pakistan might already have produced a transformation. But with military and economic support from the United States, China and several Middle Eastern states, Pakistan has been able to sustain what might otherwise well have been a hopeless rivalry with its giant neighbour.

If this analysis of Pakistan's position is correct, then the prospects for maintaining the status quo in South Asia are questionable. If the essential structure of the complex rests more on external than an internal sources of power, then its durability lies hostage to continued involvement by countries whose main interests lie outside the subcontinent.

Internal Transformation

The internal transformation of a local complex occurs when its essential structure changes *within* the context of its existing outer boundary. Such change can come about as a result either of decisive shifts in the distribution of power, or of major alterations in the pattern of hostility. The unifications of Germany and Italy during the nineteenth century, for example, constituted an internal transformation of the European security complex. The elimination of Israel as a Jewish state would mark an internal transformation of both the Arab–Israel complex and Middle Eastern supercomplex. Similarly, the fall of the white regime in South Africa would mark a transformation in the Southern African complex. The relatively simple structure of South Asia only allows three possibilities for internal transformation: the break-up of one or both of the major powers, the ending of the hostility between them, or one side achieving domination.

Within a complex, changes in the distribution of power can have the same sort of structural effects as those commonly used for describing the international system as a whole. Whatever the starting number of poles of power within a complex, change can move it either towards greater concentration or towards greater diffusion. At one extreme, is the possibility of a monopolar, or hegemonic, complex in which a single power dominates a set of minor powers. The North American complex has this character, and so would the South Asian if Pakistan either lost its external support, or disintegrated. At the other extreme is a diffuse distribution in which several more or less equal powers jostle for influence. The classical balance of power in Europe, and even more so the current power structure of the Middle Eastern supercomplex, both illustrate the multipolar pattern. Where the membership of a complex is large, prospects exist for shifts up and down from, say, tripolar to either bipolar or quadripolar. The smaller the number of principal actors, the more significant will be the shift. A move from seven to six powers might well not constitute a significant transformation, but a move from three to two definitely does.[21]

Changes in the patterns of hostility are more difficult to define, because hostility is usually harder to measure than power. One country represents only a single pole of power, but it may, as the case of Iraq illustrates, be the focus of several relations of hostility. At the extremes of the spectrum of possibilities for a local complex are chaos and security community. In chaos, each actor is the enemy of all the others. In a security community, disputes among all the members are resolved to such an extent that none fears, or prepares for, either political assault or military attack by any of the others.[22] Extreme cases are rare. In South Asia, the pattern of hostility is as simple as it could be, and the only path of internal change is towards resolution. In complexes with higher levels of polarity, there are usually core hostilities, like those centring on Iran, Israel and South Africa, which are essential to the definition of the status quo. Secondary shifts in hostility, like that between Iraq and Syria, or between Libya and Egypt, can often occur without disrupting the basic pattern of the complex. Changes in core hostilities constitute an internal transformation.

External Transformation

A security complex undergoes external transformation when its essential structure is altered by either expansion or contraction of its existing outer boundary. Minor adjustments to the boundary may not affect the essential structure significantly. The addition or deletion of major states, however, is certain to have a substantial impact on both the distribution of power and the pattern of amity and enmity. Whether or not South Asia has experienced this type of transformation as a result of the Soviet occupation of Afghanistan is a question we will consider in Chapter 9. Because it was born complete out of the process of decolonisation, South Asia did not experience an unstable formative period involving successive decolonisations of new states. Where decolonisation was less coordinated, as in Africa and the Middle East, the local security complexes underwent a long process of formation. During this process, several external transformations occurred as newly independent states joined them. The changing of the Portuguese colonies into independent states during the mid-1970s marked the last major event in this process. Other areas where potential for external transformation of the South Asian complex exists are in its relations with the Gulf and with China, and both of these possibilities will be examined in Chapter 9.

Overlay

Overlay means that one or more external powers move directly into the local complex with the effect of suppressing the indigenous security dynamic. This move can be imperial in character, and during the height of European empire, most of what is now the Third World was overlaid by direct political and military control. With decolonisation, imperial overlay has become unfashionable, its sole remaining practitioner being the Soviet Union. It has been replaced by more voluntary forms of submission, like that of Western Europe and Japan to the United States. Under this arrangement, overlay takes the form of unequal alliance. Local security concerns are overridden by the security orientation of the dominating power, and this orientation is reinforced by the stationing of that power's military forces directly within the local complex. The local states acquiesce in their own subordination either because they collectively fear some other outside power, or because they fear the further unrestrained operation of their own local security dynamic. In the case of Western Europe, both reasons apply.

The likely result of overlay is that the suppression of the local security dynamic, and/or protection against another external power, is gained at some, perhaps considerable, cost in entanglement with the larger security dynamic of external powers. Thus in Europe, the local dynamic which focused mainly on the position of Germany has been virtually dormant since the late 1940s. But in return, the old continent has been polarised into a front line in the rivalry between the United States and the Soviet Union. As the history of the NATO alliance attests, many tensions exist between the Western Europeans on the one hand, and the Americans on the other, about the nature and purpose of their joint rivalry with the Soviet Union.

In the post-colonial era, it is fair to assume that the overlay option requires exceptional circumstances to come into being. The post-war conditions in Europe and Japan provided such circumstances. The enormous upheaval of the Second World War produced not only exhaustion and weakness resulting from the self-destructive operation of the local security dynamics, but also the physical and political dominance of the United States and the Soviet Union in both areas. To be feasible, overlay requires either or both of a massive, locally applicable, military superiority by the external power(s), and a strong will on the part of the local powers to invite a large and sustained external presence. It also requires a strong enough interest on the part of the external power(s) to justify the costs of extended presence. Given the strength of feeling about national independence in the Third World, and the geographical remoteness of most of it from the centres of American and Soviet power, such conditions are difficult, if not impossible, to find in the contemporary world. The Americans discovered the cost of attempting overlay against the grain of local conditions in Vietnam. On a smaller scale, the Soviets may be learning the same lesson in Afghanistan.

The only place where superpower interest looks strong enough to lead even as far as consideration of overlay is Southwest Asia. There, Western interests in Gulf oil, and American interests in the 'third strategic zone' of containment against the Soviet Union,[23] confront a strong Soviet presence and desire to expand influence. Is the Soviet overlay of Afghanistan a precursor of larger action or an end in itself?

CONCLUSIONS

By positing a defined structure of relations at the regional level, the security complex framework raises a number of general questions. How durable are the complexes in question, and what is the relationship between their internal dynamic, and their interaction with other complexes? What role does penetration from higher level complexes play in the security of local complexes? How do the trends within the structure of security complexes influence the foreign policy options available to the states concerned? And how, in turn, do the foreign policies of individual states feed into the structure of relationships defined by the security complexes? Are all of the states concerned essentially locked into patterns of relationships over which they have little control, or do some of them have real leverage over the structure of events, and therefore real choices to make in the directions of their foreign policies?

These questions underlie all of the chapters in this volume, but will be brought into particular focus in Chapter 9, where we will reintegrate the analyses from the four levels of the framework. The seven chapters in between this one and Chapter 9 will aim to deepen and clarify the four levels. Chapters 2 and 3 will examine the domestic level within India and Pakistan, looking for the strengths and weaknesses which inform their security problem. Chapters 4 and 5 will cover the regional level, looking respectively at the rivalry between India and Pakistan, and the relationship between India and the minor states within the South Asian complex. Chapters 6 and 7 will deal with the super-regional level, concentrating on the relations between the South Asian and Middle Eastern security complexes, and the South Asian and Sino–Soviet complexes. Chapter 8 will address the global level, focusing on the Soviet–American complex as it impinges on South Asia.

NOTES

1. Zalmay Khalilzad, *Security in Southern Asia 1: The Security of Southwest Asia* (Aldershot, 1984); and Timothy George, Robert Litwak and Shahram Chubin, *Security in Southern Asia 2: India and the Great Powers* (Aldershot, 1984) unpaginated preface, both volumes.
2. See, for example: Kenneth Waltz, *Theory of International Politics* (Reading Mass., 1979); Karl Deutsch and J. David Singer, 'Multipolar Power Systems and International Stability', *World Politics*, vol. 16 (1964) pp. 390–406; James N. Rosenau, (ed.), *International Politics and Foreign Policy* (New York, 1969) chs 27, 28 and 30; V. I. Lenin, *Imperialism, The Highest Stage of Capitalism* (New York, 1939); Robert Gilpin, *War and Change in World Politics* (Cambridge, 1981); and Barry Buzan, 'Economic Structure and International Security: The Limits of the Liberal Case', *International Organization*, vol. 38, no. 4 (1984) pp. 597–624.
3. See, for example: Richard Rosecrance, *Action and Reaction in World Politicss* (Westport Conn., 1963), and *International Relations: Peace or War?* (New York, 1973).
4. Raimo Vayrynen, 'Regional Conflict Formations: An Intractable Problem of International Relations', *Journal of Peace Research*, vol. 21, no. 4 (1984) pp. 337–59.
5. Michael Haas, 'International Subsystems: Stability and Polarity', *American Political Science Review*, vol. 64, no. 1 (1970) p. 100.
6. Ibid., pp. 98–123; Michael Haas, *International Conflict* (Indianapolis, Bobbs-Merrill, 1974); and Michael Brecher, 'International Relations and Asian Studies: The Subordinate State System of Southern Asia', *World Politics*, vol. 15, no. 2 (1963) pp. 213–35. Bruce M. Russett, *International Regions and the International System* (Chicago, 1967); Louis J. Cantori and Steven L. Spiegel, *The International Politics of Regions: A Comparative Approach* (New Jersey, 1970). Brecher, with his broader concept of 'subordinate state systems' was very concerned to tackle the analytical

gulf between area and system specialists that is part of our present concern.
7. William R. Thompson, 'The Regional Subsystem: A Conceptual Explication and a Propositional Inventory', *International Studies Quarterly*, vol. 17, no. 1 (1973) pp. 89–117.
8. On this point, see Barry Buzan, 'Peace, Power, and Security: Contending Concepts in the Study of International Relations', *Journal of Peace Research*, vol. 21, no. 2 (1984) pp. 109–25. For a fuller discussion of the concept of security see: Barry Buzan, *People, States and Fear: The National Security Problem in International Relations* (Brighton, 1983).
9. Earlier, attempts to develop the idea of security complexes can be found in Buzan, *People, States and Fear* (1983) pp. 105–15 and fn 33; and Barry Buzan, 'Regional Security as a Policy Objective: the Case of South and Southwest Asia', in A. Z. Rubinstein (ed.), *The Great Game* (New York, 1983) ch. 10. These should now be considered less authoritative than the presentation given here.
10. For a recent discussion see G. S. Bhargava, *South Asian Security After Afghanistan* (Aldershot, 1983).
11. In discussion at a conference on South Asian Security, St Antony's College, Oxford, October 1984.
12. Thompson, 'The Regional Subsystem', pp. 93–101.
13. For some useful perspectives on the significance of economic factors in defining regions, see Vayrynen, 'Regional Conflict Formations', pp. 337–44.
14. George *et al.*, *India and the Great Powers*, pp. 12, 114, 142, 193–200.
15. For a fuller description of the Middle Eastern supercomplex, see Buzan in Rubinstein, *The Great Game*.
16. Stephen P. Cohen, 'Pakistan: Coping with Regional Dominance, Multiple Crises, and Great-Power Confrontations', in Raju Thomas (ed.), *The Great Power Triangle and Asian Security* (Lexington Mass., 1983) pp. 50–51; Khalilzad, *The Security of Southwest Asia*, pp. 103–5.
17. Khalilzad, *The Security of Southwest Asia*, pp. 168–78; and George *et al.*, *India and the Great Powers*, pp. 18–22, 110–23, 173–200. See also note 10.
18. George *et al.*, *India and the Great Powers*, pp. 149–234.
19. For discussion, see R. J. Barry Jones, 'Concepts and Models of Change in International Relations', in Barry Buzan and R. J. Barry Jones (eds), *Change and the Study of International Relations* (London, 1981) ch. 1.
20. Pakistan's defence expenditure has ranged between 6 and 7 per cent of its GNP since 1975, whereas India's has been steady at around, or just above, 3 per cent. See *The Military Balance 1982–83* and *1983–84* (London) respectively pp. 125 and 127.
21. For more detailed discussion on this point, see: Peter Byrd, Barry Buzan, Peter Ferdinand and William Paterson, *The Making of Foreign Policy: A Comparative Perspective* (Brighton, forthcoming 1987) ch. 2.
22. On security communities, see: K. J. Holsti, *International Politics: a Framework for Analysis* (New Jersey, 1967) ch. 16; and Karl Deutsch *et al.*, *Political Community and the North Atlantic Area* (Princeton, 1957).
23. Gary Sick, 'The Evolution of US Strategy Toward the Indian Ocean and Persian Gulf Regions', in Rubinstein, *The Great Game*, p. 79.

Part II
The Domestic Component of the Security Problem

In this part, the focus is on the domestic level of the security problem in terms of the two major states within the South Asian complex, India and Pakistan. The main concern is to find out how stable these two states are in their own right, and to what extent their governments see security in domestic rather than international terms. Looking ahead to subsequent sections, we also begin to enquire into how, and to what degree, their internal instabilities feed into their security relations both with each other and with third states.

To answer the question about domestic security, we examine the fundamental cohesion of both states as political systems. The two chapters ascertain whether the institutions of India and Pakistan rest on a coherent idea of the state, either nationalist or ideological, that provides firm foundations even for weak institutions. We look at the extent to which separatist movements, whether ethnic, religious, linguistic, or cultural in motivation, undermine the idea of the state, and/or pose threats to its institutions. We also look more generally at the strengths and weaknesses of the governing institutions, asking whether there is substantial identity or substantial alienation between the security of the government and the security of the state. Finally, we look at the character of the two national economies to find out whether they enhance or undermine the stability of government and state.

In this part we can only begin to answer the questions about linkages between the domestic level and the higher levels of the security problem, but we will lay here the foundations on which to build in subsequent sections. Issues to which information in these two chapters is of particular relevance are: the extent to which the idea of the state in India and Pakistan is not self-contained within each country, but overlaps with neighbours in some way; the extent to which separatist movements and/or political institutions connect across borders; and the degree to which economies are interdependent with those of neighbours.

2 India: The Domestic Dimensions of Security

NANCY JETLY

INTRODUCTION

It is an axiomatic truth that the real strength of a nation lies in its unity and solidarity and that national security is essentially built on the base of internal stability. Conversely, domestic instability and internal conflict, undermining as it does the government's domestic credibility, erodes the very base on which national security is built and makes the state vulnerable to external threat and manipulation. Internal factors then can at times jeopardise national security more critically than external dangers and it is in this context that domestic dimensions of national security deserve greater attention then they have so far received.[1] An unstable internal situation caused by a serious political breakdown or continued economic instability regardless of military strength becomes a crucial factor in a nation's security outlook inasmuch as it invites unwelcome external attention, which may lead to intervention by foreign powers.

Problems of nation-building and national integration are rooted in India's size and multi-ethnic society with regional, religious and linguistic diversities. Since independence, the Indian political scene has been characterised by the ongoing quest for forging links of unity and solidarity within the existing framework of diversities and cleavages. Although pressures emanating from diverse ethnic and community groups have served to deepen regional identities, and have periodically, and have periodically posed a serious threat to India's territorial and political integrity, the Indian system has been able to cope with such challenges without damaging the national cohesiveness. Although Congress's handling of the government has not been always tactful, and centre-state relations have often been strained, it has nevertheless managed to defuse the challenges emanating from religious, linguistic and regional groups claiming separate identities, within or outside the federal framework. As yet the problem of preserving India's unity has not been severe, partly, at least, because conflicts emanating from cultural pluralism are basically bargaining pressure tactics, working towards a reasonable politico-economic deal rather than secession from the national mainstream. These tactics do not necessarily impede either the process of national integration or the emergence of a democratic political community.[2]

Simultaneously, counterbalancing forces of national integration have also been at work. An increasingly integrated economy in both industrial and agricultural sectors, resulting from the horizontal spread of economic ties, has become one of the bonds helping to preserve India's integrity. This is, however, not to minimize the intensity of competing demands from different sections of society and regions for equitable distribution of developmental benefits. But a growing perception of a common stake in the viability of an integrated economy has come effectively to cut across sectional and regional demands. The thrust of a national system of communications and transportation, and of technological and scientific development, is also towards greater unity and cooperation at the national level.

LANGUAGE, RELIGION AND CASTE CONFLICTS

Primordial associations like language, religion and caste, forming as they do the core of individual or group identities and utilised as they are by the minorities to promote their occupational and political interest, continue to remain potentially divisive forces in India's national integration.

Linguistic tensions are inherent in the very existence of 15 recognised languages among 830 languages or dialects. Language riots in Bombay and Assam, violent Dravida Munetra Kazhan (DMK) agitations against Hindi as an official language, Nagaland's adoption of English as the state language, and the border dispute between Maharashtra and Mysore reflect the scope and variety of linguistic conflicts in India. However, the major challenge of linguistic regionalism in the 1950s, and 1960s, was effectively, if rather belatedly, blunted by the linguistic reorganisation of states in 1956. The Punjab was, however, reorganised on a linguistic basis as late as 1966 with its split into the states of Haryana and the Punjab. Although major problems of linguistic reorganisation of the states has been accomplished, the far-reaching implications of such division for the future of Indian unity, or the question of the status and rights of linguistic minorities in various states, have yet to be resolved. The latter problem is quite acute as may be seen from the recent Bengali–Assamese tensions in Assam or the continuing irritation among the Muslims over the denial of official status to Urdu.[3]

The controversy over the official language, which at one point threatened to split the nation on North–South lines, was also gradually defused by the formulation of the three language formula. Following sustained and at times violent opposition to the abolition of English as the official language in favour of Hindi in 1965, as envisaged by the Constitution, the government shrewdly accepted the continued use of English as long as the non-Hindi states wanted it. In effect, it is now a 'two language policy for official transaction ... three language policy for the school system and a regional language policy for Union Public Service examination.'[4] Such a flexible arrangement was unavoidable because any premature attempt at unification in linguistic terms would have only heightened the regional grievances and posed a threat to the nation's solidarity.[5] It may be pointed out here that although the potential threat of linguistic separatism crystallising into political antagonism in the regional context remains, fears of national disintegration have been eased by the gradual, though imperceptible, dissipation of resistance to Hindi as a result of constitutionally guaranteed usage of English.

India is constitutionally a secular state but as Jawaharlal Nehru pointed out, the fact has yet to be reflected in India's 'mass living and thinking'.[6] Although no discrimination is practised on religious grounds, religion has yet to be displaced from the formal as well as informal arenas of politics in India. Relations between the Hindu majority and Muslim minority form the core of the communal problem in India. The seemingly intractable nature of Sikh demands in the Punjab has added a new dimension to the problem and its consequences for India's secularism are yet to unfold. Hindu–Christian relations have been generally marked by an absence of tension and violence, notwithstanding certain recent stray incidents in the south.

Communal tensions, sporadically leading to violent riots often on flimsy grounds, have become a familiar feature of the Indian system. But these tensions, although a familiar feature, seem to be increasingly sustained not by cultural and religious differences but by the structure and pattern of political mobilisation – relying heavily as it does on communal, caste and religious networks – which in turn operates on a specific economic situation.[7] A certain basic reorientation of communal relations seems to have taken place, following the changes in the traditional stratification between Hindus and Muslims as a result of a slight economic betterment of Muslims. It has been generally the beneficiaries of education and economic development who have been the most vocal and assertive in their demands. As such, communal tensions today are a natural process of purely secular changes of economic and social development, rather than an indication of the breakdown of the system, and these tensions thus do not necessarily constitute 'a distinct category of tensions to be distinguished from the other forms of general civil strife occurring from time to time within Indian society'.[8]

Manipulation of religious sentiments by political parties – even the professedly secular ones – has, however, been established beyond doubt. Politicians have vested interests in fomenting insecurity among Muslims – through accentuation of Hindu–Muslim differences – to ensure the Muslim bloc vote for electoral gains. The Congress Government itself has unashamedly followed the policy of 'political assimilation and cultural pluralism' in order to gain the Muslim bloc vote without entailing any significant loss of Hindu votes. This has involved accommodation on cultural demands like encouragement of Urdu and the preservation of Muslim Personal Law on the one hand, and resistance to political demands, like proportional representation and formation of religious parties, on the other.[9]

Another contributing factor in the perpetuation of communal tensions is the role of administrators. The minority group sees itself as being at the mercy of civil institutions which, following official differentiation between state and religion, are dominated by the partisan Hindu majority. Communal organisations like Jaamat-E-Islami and Rashtriya Sangstha Sangha (RSS) have also exploited the ideology of communalism for narrow gains. While the former plays on the minorities' grievances, perceived or real, about unequal access to economic and political power, the latter stresses 'Indianisation' of the minorities. This only serves to compound the minorities' fear about Hindu communalism, the 'major threat that secularism in India has continued to face since independence'.[10] The grievances range all the way from inadequate representation in administrative and political echelons to denial of official status to Urdu and efforts at legal infringement of Muslim Personal Law. Recently, the revivalist posture of Vishwa Hindu Parishad, professedly a non-political religious organisation, has raised spectres of an intensified Hindu–Muslim confrontation.[11]

However, the growing identification of Muslims with the national mainstream, with its attendant political advantages and economic benefits, has given them an increasing stake in the Indian polity. Their geographical dispersal (except in Jammu and Kashmir where they are in a majority) also hinders their capacity for more effective political bargaining and to that extent blunts their potency as a disintegrating force. However, Hindu–Muslim tensions, although within manageable proportions, continue to pose a major dilemma for India's secularism and integration. The Muslims themselves do not harbour any secessionist plans, but insensitive handling of the Muslims, as in Kashmir, could easily push them into an insurgency comparable to that of the Sikhs in the Punjab.

Caste loyalties also remains a potentially disintegrating force, but these have over the years ceased to pose the same kind of threat to the Indian nation-state as the religious, linguistic or tribal loyalties. Caste, while losing some of its social rigidities has increasingly become a political force, playing an important role in the functioning of Indian polity. Social stratification has tended to yield place to political stratification, transforming the near exclusive role of caste in the social set-up to a conjunctional role in democratic politics. Constitutional safeguards for the backward sections have in fact given a new political lease to the caste system. Initially introduced to protect the interest of the downtrodden classes, these are now extended to consolidate political advantages or reap electoral dividends. Although no political party functions from a single caste base and no caste votes *en masse* for any single party, certain identifiable patterns of caste–party equation have come to stay. For instance, over the years a special relationship has existed between the Scheduled Castes and the Congress Party.[12] At the same time notice must also be taken of the fact that political alignment often cuts across caste identity, and thus secularising it in the process.[13] Political necessity has made for economic and social cooperation among caste groups along lines of common class interests. These cut across caste boundaries, and enable them to bargain with a political party or government for more economic benefits and privileges.

Caste identities and ties, however, remain important vehicles for mobilising political and electoral support within the present political system. The sheer numerical force of backward classes makes them an important factor in the contemporary Indian polity, underpinned as it is by adult franchise, making them a target of manipulation by parties of all shades. Growing identification with the national mainstream through democratic channels gives them a further stake in the political system and nation-building processes. This has, however, meant no appreciable abatement in the atrocities on, and exploitation of, the Harijans, Incidence of caste violence is on the increase in many states particularly in Uttar Pradesh, Bihar and Haryana. Recently there have also been disturbing reports of a backlash from the educated but economically deprived middle classes, culminating in unprecedented large-scale violence in Gujarat, following a protest for scrapping of all reservations for Harijans and tribals in the post-graduate medical courses. On the other hand, various movements championing the cause of 'forward community' for reservation of jobs and seats in educational institutions on the basis of economic backwardness have sprung up in states like Bihar, Kerala and Tamil Nadu.

Caste rivalries continue to play an important role at the local and state level – affecting voting behaviour, distribution of tickets and even ministerial appointments – but are less effective at the national level. For instance, the Kamma–Reddy rivalry has for long decisively dominated Andhra Pradesh politics. Similarly, no political party in Tamil Nadu can relax its anti-Brahmin posture for fear of losing its grip on the state politics. On the whole, however, as caste tensions based on the concept of purity and pollution have increasingly acquired the nature of power tensions – subjected to secular and economic forces and related to political power, these have got accomodated within the acceptable parameters of the Indian political system. To that extent, the potent threat of caste loyalties as disintegrative forces has been considerably defused.

THE WEAKNESS OF POLITICAL INSTITUTIONS

The Indian political scene today is marked by an uncertainty following the gradual erosion of Congress (I), the ruling party, and the dim prospects for the emergence of a credible alternative to it.

It must, however, be made clear at the outset that Congress (I) remains the only genuinely national party on the Indian political scene today, notwithstanding the fact that over the years it has become increasingly torn by factionalism.[14] There is practically no state today where dissidents are not challenging the incumbent chief minister on charges of inefficiency or corruption, or both. Gundu Rao's style of functioning was no insignificant factor in the Congress (I) defeat by the Janata-led front in the 1983 Karnataka Assembly elections. At times, Mrs Gandhi's efforts to enforce discipline by suspending some local dissidents were forestalled by the dissidents themselves forming a separate party. For instance, in Gujarat Congress dissidents formed a separate 'Rashtriya Congress Party'. Paradoxically, Mrs Gandhi's unprecedented centralisation of power had only exacerbated factionalism and led to progressive growth of dissidence within the party at the state level. The gradual erosion of the institutionalised patterns of political power was largely the fall-out of Mrs Gandhi's authoritarian style of functioning since the 1969 split.[15] Thus, while, on the one hand, induction of personally loyal chief ministers had resulted in the extraordinary concentration of power with Mrs Gandhi, on the other hand, the centralisation of power had made her wary of the emergence of any leaders with an independent and stable base. While the political functioning of the Congress in the Nehru years was generally marked by stability of the state units and relative decentralisation within the party, the federal system under Mrs Gandhi had been reduced to 'one woman, one government' rule in which the chief ministers retained office only at the whim of the leader.[16] Indefinite postponement of elections in the party organisations since 1972 has not only systematically eroded the intra-party democracy, but has also made the state party organisations and governments increasingly subservient to, and dependent on, the centre.[17].

Factional infighting which has become an endemic feature of the ruling party's politics threatens to wear out the very base of popular support for the party. A series of quick changes in the chief ministers of states like Maharashtra, Andhra Pradesh and Gujarat met with little success as far as factionalism was concerned. In fact, the farce of centrally imposed chief ministers (three changes in 1982) became a major plank in the victory of Telugu Desam – a rank outsider on the party scene, in the 1983 elections in Andhra Pradesh. Factionalism had also cost the Congress (I) some seats in the Rajya Sabha elections in March 1982, because of cross-voting by dissidents from Andhra Pradesh and Maharashtra. It also failed to emerge as the majority party in the mid-term elections in both Haryana and Himachal Pradesh.

The national opposition, comprising an assortment of national regional and personal parties,[18] is in no less of a disarray and despite moves and countermoves for unity has yet to provide a credible alternative to the ruling Congress. Efforts at building any United Front without the participation of the Left partners and the Bharatiya Janata Party (BJP) would hold little promise of providing a viable alternative to the Congress (I). Earlier efforts in such a direction resulting in the formation of the National Democratic Alliance, Bharatiya Janata Party (BJP) and Lok Dal, in August 1983, and United Front (Janata, Congress (S), and some other parties) have also come to nought. After a brief break in 1977–9 which witnessed both the emergence and disastrous collapse of an alternative government, the 1980 elections have restored India to its normal political state characterised by the Congress preponderance and a fragmented opposition.[19] The landslide victory of the Congress (I) in December 1984 confirmed that pattern.

Joint efforts at electoral understanding notwithstanding, any sustained cooperation among the opposition parties – beset as they are by personality and policy divisions – remains a distant dream.[20] This would make for continued political instability in India. The four opposition conclaves since May 1983, while underscoring the general willingness of the opposition parties to close their ranks to challenge the Congress (I) supremacy, have remained meaningless exercises in so far as crystallisation of any joint constructive policy or programme is concerned. The limited draw of the conclaves was underscored by their failure to involve parties like the Lok Dal and Bharatiya Janata Party. Overall, the Indian political system remains characterised by a pitifully fragmented party structure which only serves to underline the necessity of political coalitions to win elections and form governments, but does not impart any stability to the system.[21]

CENTRE–STATE RELATIONSHIP

The incipient threats to the Indian federal polity also make for growing strains on the Indian political structure. For instance, regionalism in India which was basically a function of cultural and linguistic diversity has now also acquired increasingly political and economic dimensions. Regionalism today is expressed in demands for greater state autonomy, Punjab and Assam representing the extreme instances of such aspirations. The continued instability in the northeast involving as it does basic interests of national security, is yet another dimension of the growing strains on the Indian political structure.

Strains in centre-state relations flow partly from the inbuilt centrist bias in the Indian Constitution and partly from the functioning of the political processes. The legislative, administrative and financial arrangements are heavily weighted in favour of the central government. The residuary power vested in the centre for transfer of functions from the State list to the Union list, and administrative clout in the use of emergency powers by the centre to impose President's rule, and the immense fiscal powers of the centre in appropriating to itself resource yielding powers, have all strengthened the centre's claims over the states. Some analysts have termed the central government as the 'government of the governments in India'.[22] The course of national planning since 1950, the creation of All India Services and a long spell of Congress rule both at the centre and the states has only served to reinforce the unitarian tendencies of the Indian federal set-up. Questions of centre–state relationship remained generally dormant in the first two decades of independence as these, under the Congress dominated system, came to operate along the party lines, the degree and modalities depending on the political stature of state leaders and their relation with the central leadership. However, even in the Nehru era of 'consensual' politics, Nehru had set definite limits to the centre's willingness to accommodate dissent on the part of the chief ministers. Although Nehru may not have initiated efforts to topple the non-Congress governments, he certainly did not hesitate to use the federal machinery, especially presidential rule, to further the interests of the Congress organisation.[23] The advent of non-congress ministries in a number of states – ranging all the way from left to right – following the fourth general elections in 1967 witnessed the first audible rumble of discontent with the existing framework of centre–state relations and a demand for larger autonomy for the states. While the two communist parties championed wider economic and financial powers for the states, Tamil Nadu – the only state to have returned a purely regional party to power – openly accused the centre of discrimination. It appointed the Rajamannar Committee in 1969 to go into the whole gamut of centre–state relations. The committee, among other things, recommended the restriction of the scope of Article 356 and transfer of certain items from the concurrent list to the state list and vesting of residuary powers of legislation and taxation in the state legislatures. The re-emergence of the Congress both at the centre and the state in 1971 resulted in concentrated moves in the reverse direction inasmuch as the dominance of central political and bureaucratic structure over the states was exerted more effectively by Mrs Gandhi's government. The imposition of emergency in 1975 further lent a predominantly centrist orientation to the federal structure. During the years 1966–77, out of a total 39 cases of intervention in the states through Article 356, as many as 23 pertained to the non-Congress governments.

The advent of Janata Government in 1977 – the first non-Congress government at the centre since independence – witnessed some attempt at the restoration of 'federal values' in response to renewed demands for political and economic autonomy from the states. In actual practice, however, the Janata government's initiative remained limited to abstention from the exercise of over-centralisation, and allocation of slightly greater weight to the states in the decision-making process: it initiated no fundamental moves to correct the imbalances by restructuring the existing financial and administrative arrangements. The Janata phase witnessed not only partisan use of central powers by dissolving nine non-Congress state ministries at one go through a presidential proclamation, but also a systematic interference in the Janata-ruled states in pursuance of the power struggle raging at the centre. Janata chief ministers became the victims of the power struggle between Bhartiya Lok Dal (BLD) and the Jana Sangh. BLD, in particular chief ministers, were repeatedly asked to seek a vote of confidence from their legislative parties.[24]

The centre–state debate acquired an urgency after 1980 with the comeback of Mrs Gandhi to power on the one hand, and the simultaneous stabilisation of the existing non-Congress governments in West Bengal, Tamil Nadu and Jammu and Kashmir, also the emergence of new non-Congress governments in Andhra Pradesh and Karnataka, on the other. The unresolved crises in the two sensitive states of Assam and the Punjab have brought the centre–state confrontation into sharper focus. The recent conclaves of some non-Congress chief ministers are pointers towards crystallisation of a hardline approach towards demands for greater political and economic autonomy. It is in this context that the governor's role and the misuse of Article 356 has come increasingly under severe criticism. Besides repeating the Janata tactic of desolving nine non-Congress state governments, Mrs Gandhi's cynical use of Article 356 in Sikkim, Jammu and Kashmir and Andhra Pradesh (revoked in the last state under persistent public pressure) only aggravated the existing tensions. Another important element in the controversy relates to the demand for the reassessment of fiscal arrangements between the centre and states, in view of the centre's vast financial powers and the pressing developmental needs of the states. The state governments themselves, however, leave much to be desired in respect of the management of resources already at their command, as reflected in the functioning of their planning bodies, ongoing projects and overall integrated development programmes. Fears are also voiced about centre–state relations turning into 'an anti-centre tirade', partly to cover up the inadequacies and failures of the individual state themselves and this tends to take away much of the force of the states' tirade against the centre. Also economic demands voiced by the states vary in their scope and emphasis. While Tamil Nadu and West Bengal want an enlarged share of revenue, Jammu and Kashmir want the continuation of existing fiscal arrangements to protect the interests of economically advantaged states.

The appointment of the Sarkaria Commission in 1981 to recommend appropriate changes in centre–state relations within the present constitutional framework was both a device to preempt the states' growing demands, and a timely recognition of the dangerous implications of explosive situations in the Punjab and Assam for centre–state relations. The government has not accepted – even in part – recommendations of the earlier committees (including the ones set up by itself like the Administrative Reforms Commission 1967). The fate of the Sarkaria Report remains uncertain. Although there is no inherent contradiction between a strong centre and a powerful state, centre–state relations in India are essentially a function of political dynamics. Their future therefore depends more on sensitive political handling than on mere restructuring of constitutional arrangements.

THE THREATS OF SECESSION

The continuing civil strife and incipient secession in the strategic north-east poses a grave problem for India's security. Ever since independence, an uneasy peace has characterised the central government's relations with the hill areas in the north-east which have been compounded by the region's poverty, relative remoteness and cultural alienation. The Indian government's extension of administration to the outlying tribal areas – under loose control during British rule was largely perceived by the indigenous population as a ploy to destroy their tribal culture and traditions. Continuing tensions between the Assamese plainsmen and most of the hill tribes further intensified the latter's fears of economic exploitation and cultural and religious domination. The Nagas and the Mizos even resorted to full-scale armed insurrections for secession from India. Following the creation of Nagaland in 1963 and the formation of Mizoram in pursuance of the reorganisation of the north-east in 1972,[25] insurgency in the area has to a large extent been brought under control, but peace has yet to come to this region.

The complexity of the unsettled situation in the north-east is a product of deep rooted political and economic factors. Crises of identity, rising unemployment among the educated youth, transborder ethnic links, and the machinations of foreign powers not only sustain the revolutionary fervour in the area but also make for generally fragile links with India. This has necessitated a visible army presence in practically the whole of the north-east. Tripura, is the only state in the north-east not to be brought under the Disturbed Areas Act. There are reports of the centre pressing the CP(M)-led government in Tripura to allow the introduction of the army in view of the sustained terrorism indulged in by Tripura National Volunteers which has formed a government in exile, the 'Free Tripura' in Bangladesh territory.[26]

Insurgencies continue to be a part of the political scene in Manipur, Mizoram and Nagaland. In Manipur, the People's Liberation Army (PLA) – a product of Meetei nationalism and communist ideology – is working for the liberation of the Manipur Valley. Although Chinese help to the Mizo rebels has been stopped, there are reports of their cooperation with the White Flag rebels in Burma. The Mizo National Front (MNF), outlawed since 1979, has intensified its activities to achieve an independent Mizoram.[27] This coalition, despite a recent report of a unilateral ceasefire by Mizo rebels under instructions from Laldenga, which might have paved the way for resumption of dialogue with the central government to end the twenty year-long insurgency in Mizoram. But in view of past experience, the outcome does not appear hopeful. There are also, on the other hand, reports of growing links between the Mizo National Front and the Tripura National Volunteer extremists, creating a highly volatile situation in Tripura.

The formation of the Seven United Liberation Army (SULA) in 1980 by the underground rebels in Assam, Nagaland, Manipur, Mizoram, Tripura, Arunachal Pradesh and Meghalaya, is a cause for further concern, signifying as it does a concerted move towards organised secessionist insurgency in the region. The geopolitical vulnerability of the north-east – ringed as it is by China, Burma, Bangladesh and Bhutan – has far-reaching implications for Indian security. Although China no longer provides arms and training to the rebels, Burma, having no effective control over its northern areas, continues to be a source of arms and sanctuaries for the insurgents. There are also suggestions of Bangladesh's involvement, albeit indirect, in the insurgency movements. Despite sustained counter-insurgency operations by Indian security forces, not much success has been achieved. A long-term solution in the region would have to be political. Meanwhile, the unsettled conditions in the whole of the north-east remain a source of concern to New Delhi.

The agitation in Assam was carried out largely under the leadership of the students organisation All Assam Student's Union (AASU) and All Assam Gana Sangram Parishad (AAGSP). These comprise a collection of political and non-political organisations under a common banner, but their activity has lost some of its initial momentum and the state seems to have settled to an uneasy calm. The Assam question, with no resolution of the problem in sight, remains fraught with dangers which may have dangerous implications for India's national integrity. The strategic geophysical position of Assam adds yet another dimension to the intractable problem.

Triggered by an almost casual discovery of the 'foreigners' in the 1979 electoral rolls, the movement went on to encompass virtually the entire Assamese-speaking population in a serious confrontation with the centre.[28] The agitation, centred around the detection and deletion from the existing electoral rolls of all 'foreign' settlers as per existing constitutional laws also demands not only the stopping of further illegal entry of foreigners but also their explusion from Assam.[29] The core of the problem lies in the large-scale influx of Bengali immigrants into the state which threatens to reduce the Assamese to a minority in their own homeland. During the years 1901–71, the population of Assam registered a growth rate of 600 per cent as against 130 per cent for India as a whole. In a report in 1980, the Chief Election Commissioner warned that in view of the continuing influx from neighbouring areas, by 1991, foreign nationals would form a sizeable percentage, 'if not a majority of the population of the state'.[30] The massive influx of non-Assamese has led to increasing alienation of land, loss of employment opportunities and economic hardships for the Assamese population. Assam also perceives itself as a victim of the centre's neglect in terms of development, communications and education.[31] No major industries have come up in Assam since independence and the spread of communications within the state leaves much to be desired.

The mass upsurge against foreign nationals in Assam cannot, however, be viewed in isolation from the larger question of Assamese nationality. Assamese élites have been known to draw little distinction between foreigners and outsiders. Agitations have become an endemic feature of Assam politics and the present agitation is not the first of its kind. Anti-official language agitations (1955, 1958–60), anti-Marwari agitation (1966) and anti-government agitation over its plan for a federal set-up for the undivided Assam are instances in point.[32] Assamese élites – conscious of a distinct linguistic, cultural and political identity – have not really been able to handle the question of Assamese identity within the larger national identity. The present movement has thus laid itself open to the charge from various quarters of being anti-national, anti-Muslim and even secessionist. The West Bengal Assembly unanimously passed a resolution in 1980, characterising the movement as separatist and secessionist inasmuch as it went against the right to security of minorities in every state.[33] The movement is thus seen as a revival of the age-old campaign against Bengalis and non-Assamese in a new garb, with virtually no difference made between the Bengali-speaking Indians and the Bengali-speaking Bangladeshis. The conflict has also increasingly acquired communal overtones, with both Hindu and Muslim fundamentalist organisations fishing in the troubled waters.[34]

Tribal unrest in Assam against both the Bengalis and the Assamese adds another dimension to the Assam problem, as evidenced by the Nellie massacres in 1983. The Plains Tribal Council of Assam is even claiming a separate tribal state of Udayachal to be carved out of Assam.[35]

Although the government has assured the Assamese of the preservation of their social, cultural and linguistic identity,[36] it has achieved little by way of a long-term solution of the problem. Negotiations between the government and the Assamese leaders have floundered on the question of the 'cut-off' year for the grant of legal residence to the 'foreigners'. The Assamese leaders have been pressing for 1961 as the 'cut-off' year while the government insists on 1971. The situation has become more complicated in view of the recent Supreme Court judgement validating the 1979 rolls and the Election Commission's suggestion for retaining the 1971 electoral rolls as the basis for the next elections. Although after the large-scale violence which rocked Assam following the promulgation of the

Essential Services Protection Ordinance and Enforcement of Assam Disturbed Area Act (March 1980), a semblance of normality has returned after the restoration of the popular government in 1983. But no real solution to the Assam problem is yet in sight. The mid-term elections in Assam were largely farcical, only 12 out of 14 seats for the Lok Sabha could be filled due to widespread violence.[37] Even an agreement on the cut-off year would still leave unresolved the major problems of deportation. Given the constitutional constraints, the prohibiting of any resettlement – voluntary or otherwise – of immigrants outside Assam remains slight. The deportation of the Muslims would not only lead to communal violence, but would also strain India's relationship with the touchy Bangladeshis. Any further delay in settlement might well lead to the growth of a secessionist movement. The government's 'holding operations' may not hold too long; on the other hand, any durable settlement, based as it would have to be on constitutional safeguards guaranteeing the Assamese linguistic, cultural and political identity, would raise larger questions regarding the existing framework of centre–state relations in general. And therein lies the dilemma of the government's response to the Assam problem, which continues to gnaw into the vitals of the country.

The recent happenings in the Punjab – the sustained violence of the extremists and the government's heavy-handed response – poses by far the most serious problems for India's integrity. The government's recourse to army action in June 1984 – Operation Blue Star – following the inability of the police and para-military forces to check increasing militancy in the state has raised doubts about India's national unity. The present Punjab imbroglio – an explosive mix of language, religion and politics – does not stem from any immediate factors, but is deeply rooted in the past history of the Sikhs and the state, which is centred around the Sikh quest for a separate identity because of fears of Hindu domination and eventual absorption.[38]

The failure of the Akalis – the party generally identified with the Sikh aspirations to capture power in the state single-handedly and retain it uninterruptedly – has led it gradually to voice demands for greater autonomy for the state in a bid to ensure Sikh hegemony in the Punjab. The precarious majority of Sikhs of 2 per cent over the Hindus in the Punjab does not get reflected in the voting structure due to caste dissensions among the Sikhs – the non-Jats have by and large voted with the Congress. The Anandpur resolution, which has three different versions in circulation, has among other things, urged for greater state power; the limitations of central intervention to matters of defence, foreign affairs, post and telegraph, currency and railways, and readjustment of the state's boundaries. The territorial claims included some demands on Rajasthan and Himachal territory, and insistence on reopening the river water agreement with Rajasthan. No section of the Akali Dal – moderate or extremist – has repudiated the Anandpur resolution despite the institution of the Sarkaria Commission or the government's offer to refer the territorial demands to a Commission or adopt any other alternative acceptable to both Haryana and the Punjab.[39] This resolution also for the first time described the Sikhs as a 'nation', and was later endorsed by the Central Gurdwara Management Committee (SGPC) and the Akali Dal (both the Harchand Singh Longwal and Tohra factions). However, none of the Akali Dal factions except the miniscule group led by Sukhjinder Singh has defined Sikh nationalism in terms of a sovereign Sikh state.[40] The central government has, however, chosen to view Akali ambiguity on the concept of 'Sikh separatism' as a 'cover for subversive and anti-national forces to operate in the secure knowledge that they would not be politically disowned'.[41]

By and large, however, the extremists' demand for Khalistan notwithstanding Akali demands, would have to be looked at in the context of seeking accommodation within the framework of Indian unity. The demand for Khalistan was first put forward by certain members of the Sikh communities in the United Kingdom and the United States. Later, it found some response from some extremist groups on the fringes, notably Dal Khalsa, under the broad patronage of Sant Bhindranwale. It would, however, be impossible to term all Sikhs as Akalis and all Akalis as separatist militants just because the central demands of the Akali Dal – which could rationally be the demands of all Punjabis – have got lost in the maze of militant religious demands and Sikh grievances.[42]

It would, however, seem to be more than a coincidence that rapid escalation of the crisis – through a series of agitations culminating in the launching of 'Dharamyudh' in 1983 – should have synchronised with the Akali Dal's ministry's dismissal in 1980, with no immediate chance of its gaining power through electoral processes. The simultaneous growth of a Sikh fundamentalist movement under Sant Bhindranwale since 1980 changed the contours of the movement from a party agitation to a Sikh movement encompassing divergent orientations and rationales, further diminishing the prospects of a negotiated settlement. There are reasons to suspect the Congress (I) role in the build-up of Bhindranwale as a counter-force to growing

Akali strength. As it happened, Bhindranwale was to emerge as an independent centre of power in the Punjab with disastrous results. The gradual immobilisation of the Akali Dal, following the erosion of its credibility and the emergence of Bhindranwale and his group as the focal point of Sikh aspirations, coupled with the near paralysis of practically the entire state administration resulted in an explosive situation in the Punjab, eventually leading to the deployment of the army. By failing to take prompt action against Bhindranwale and religious fanaticism, the government deliberately allowed the situation to escalate merely for narrow electoral advantages.[43] The army action – generally greeted with relief – has been criticised on the grounds of the timing and the operational fiasco rather than the rationale of the action. Of course, repeated induction of the army for civilian duties may auger ill for the largely apolitical character of the Indian army as also the traditional acceptance of broad civilian control. Developments in neighbouring countries tend to reinforce such fears.[44]

The government's claim of foreign powers' subversive activities towards undermining India's national solidarity is not entirely convincing. The problem nevertheless acquires added urgency in view of the fact that growing alienation between the army and civil population of the Punjab (which borders on Pakistan) may lead to a weakening of the latter's traditional feeling of solidarity with the armed forces, which has been so far a valuable supportive asset in times of foreign aggression.

The crisis in the Punjab is confounded by the growing alienation between the Hindus and the Sikhs. This has echoes in the economic sphere as well where the Hindus have traditionally dominated trade and marketing. The deepening economic crisis in the state sharpens the cleavages. Sikhs have also perceived, probably unjustly, discrimination in representation in services and professions. The need for communal harmony in the state is underscored by the fact that the two communities are too evenly matched to allow any unbridled assertion of power on either's part. Any concession to overtly religious demands or to Sikh misuse of religious symbols for antinational activities would have a dangerous portent for similar manipulation by other religious minorities. It is in this context that the talk of Hindu 'backlash' and 'resurgence' in the Punjab foreshadows long trouble in this sensitive frontier state.[45]

Rajiv Gandhi's government has accorded a high priority to the Punjab crisis, but despite numerous concessions to the Sikhs, the initiative still remains with the extremists. Although Rajiv is both sincere and aware of the dangers of continuing unrest in the sensitive state of Punjab on the borders of Pakistan, he is constrained from taking bold action because of fears of alienating his Hindu supporters. The result is that the concessions, even though quite substantial, are made piecemeal, grudgingly and lack unanimity. Even the legitimate Sikh demand for an enquiry into the killing of Sikhs following Mrs Gandhi's assassination was not conceded until after six months and then only under threat from the Sikhs. The problem calls for a bold political solution. The majority of the Sikhs are not secessionists but merely pushed into extreme positions by the government's continued insensitivity about the centre–state relationship. While the situation remains grave, a political resolution of the crisis is distinctly within the realms of possibility.

ECONOMIC FACTORS IN DOMESTIC SECURITY

As for economic factors, which to a great extent underpin a country's stability, an increasingly integrated economy, despite regional disparities, has become one of the most effective bonds for Indian unity. India's economic development, in purely statistical terms, has been fairly impressive, as reflected in the expansion and diversification of its industrial capacity. India currently ranks at the tenth industrial nation in the world and has the third largest pool of skilled and trained manpower. Indian GDP (Gross Domestic Product) has registered an annual increase of around 4 per cent during the sixth plan (1980–85). Both agriculture and industry have shown substantial growth: in 1981 the rate of growth was 4 and 8 per cent respectively. The approach paper to the seventh plan (1985–90) targets an annual growth rate of 5 per cent, with corresponding targets of 4 per cent growth in agricultural products and 7 per cent in the industrial sector. Investment rate is expected to go up to 26 per cent of GNP as against 24.4 per cent during the sixth plan.[46] The rate of inflation has also registered an appreciable decline, coming down from a galloping 22 per cent in 1979 to a manageable 10 per cent in 1981. India has also acquired self-sufficiency in food grains, its production touching a record yield of 12.98 crore tonnes in 1980–81. In 1981, India imported food grains – the first time since 1976 – mainly to hold the price line and augment its buffer stocks. Many experts, however, feel that the food output has reached a plateau and higher growth, though attainable, is not desirable in terms of cost–benefit ratio.[47]

This rosy picture of the Indian economy, however, begins to pale when one looks at the growing rate of unemployment, continuing population explosion, incidence of a parallel black economy and the steadily rising number of people living below the poverty line. A recent study by the Federation of Indian Chambers of Commerce and Industry (FICCI) projects the number of unemployed to touch a peak of 46.26 million at the end of the current plan period. The total number of people below the poverty line in the year 2000 is estimated at 394 million, or more than India's population at the time of independence.[48] Failure to implement the land reforms – whether due to lack of political will or administrative initiative – compounds the acute problem of poverty in rural areas, notwithstanding some steps at redressal in the course of the sixth plan.

India also continues to face an acute balance of payment deficit, persisting largely due to dependence on oil imports, but the situation is likely to ease as India steps up its own oil production. The massive International Monetary Fund (IMF) loan of £5.8 billion, although providing significant short-term relief, cannot provide an effective remedy for an unfavourable balance of trade. By the mid-1980s, debt service and repatriation of the loan, coupled with loans from other sources is expected to move up to a staggering figure.[49] Growing aid-weariness and rising protectionism in the developed countries do not hold favourable prospects for rapid industrial growth needed desperately by India to bridge the gap.

The lacklustre performance of the public sector in critical areas (cement, power, transport, oil and steel), growing labour unrest in the industry, increasing militant farmers' movements, high rates of inflation and inequality in the distribution of income do not provide cause for an optimistic outlook either.

The present political scene presents a no less disturbing picture. The increasingly centralised and repressive character of the government and the consequent vulnerability of the state apparatus leading to a process of fragmentation at all levels causes anxiety.[50] Gradual erosion in general moral standards, progressive collapse of political and public institutions,[51] social unrest as evidenced by anti-reservation riots or incidence of caste violence, virtual collapse of large segments of the educational system, poor quality of leadership and widening economic disparities all go to underline the basic national malaise which may well foreshadow a serious breakdown of the Indian polity in the years to come. However, Rajiv Gandhi's most impressive electoral victory in the general elections of December 1984 and his determined efforts to revamp and reorganise the Congress (I) into an effective party organisation give reasons for optimism. But whether he will be able to check the erosion of institutions remains to be seen.

CONCLUSIONS

Notwithstanding the existence of grounds for such apprehensions, the Indian democratic polity continues to demonstrate remarkable resilience in its functioning, and the performance of the Indian economy belies the scepticism of its worst critics. Also, despite the undeniable pressure of fissiparous social and political cleavages, India's national integrity remains fundamentally secure. The challenges emanating from caste dissensions, communal tensions and linguistic diversities have been effectively managed within the existing framework, without causing much damage to the basic national cohesiveness. This assessment would remain largely valid notwithstanding the growing incidence of caste violence, communal riots or 'sons of the soil' movements in some parts of the country.[52]

The growing clamour for greater state autonomy may continue to be at times compounded by turbulent politics as in the sensitive border states of the Punjab and Assam or by chronic insurgency as in the north-east. Although these will continue to be a cause for concern, as of now they remain well within manageable limits: the coercive power of the state is sufficient to check such tendencies. No separatist movement of any great significance has yet emerged on the scene to pose a really serious threat to the nation's integrity. An important point which one should not underestimate is the political leaders' stake in not pushing separatist and secessionist demands to one for autonomy within the framework of Indian unity and integrity following its rise to power in the state amply illustrates this. This is not to minimise the danger of elemental forces overtaking the political leadership as happened recently in the Punjab. This only shows that some problems, if left unresolved indefinitely, can acquire dangerous proportions and have far-reaching implications for the future.

Serious erosion of the 'Congress system',[53] which had in the first two decades of independence served as an umbrella for varying interests and ideologies, plus a virtually non-existent national opposition and rampant factionalism at all levels no doubt causes worry. Challenges from new regional groups are sought to be blunted by recourse to indiscriminate, almost reckless, destabilisation of non-congress governments in Jammu and Kashmir, Andhra Pradesh or Karnataka, further crystallising the patterns of instability in the political system. But what is more important is that Indian democracy remains a functioning democracy despite certain undeniable distortions. The General Elections of 1977, 1980 and 1984 underline not only the national consensus on the continuing legitimacy of the parliamentary system, but also the people's decisive assertion of their political rights. The recent success of Telegu Desan in Andhra Pradesh reinforces this trend.

The Congress (I), despite organisational disarray remains the only effective governing party. As long as it enjoys a nation-wide base, and demonstrates the capability to effectively manage its factional disputes, it will retain the key to political stability in India. Any electoral or political failure of the Congress, without any credible alternative to fill the vacuum, would put a more severe strain on the polity than is discernible today. This is, however, not to underestimate the incipient challenges from systematic erosion of norms and institutions that threatens to undermine the very foundations of the political system. The need of the hour is to reconcile the existing diversities of language, culture and religion with a unified national consciousness. There is ample cause for concern but little for despair.

NOTES

1. Two recent authoritative studies on Indian security, for instance, take little note of domestic dimensions. K. Subrahmanyam's *Indian Security Perspectives* (New Delhi, 1982) and U. S. Vajpai's *Indian Security: The Politico-Strategic Environment* (New Delhi, 1984).
2. Iqbal Narain, 'Cultural Pluralism, National Integration and Democracy in India', *Asian Survey* (Berkeley: California) vol. xvi, 1976, p. 903; Paul F. Brass, *Language, Religion and Politics in North India* (Delhi, 1975) pp. 12–4.
3. See Brass, *Language, Religion and Politics in Northern India*, pp. 21–2.
4. Jyotindra Das Gupta, *Language, Conflict and National Development* (Bombay, 1970) p. 259.

5. Ibid., p. 270.
6. S. Gopal (ed.), *Jawaharlal Nehru: An Anthology* (New Delhi, 1980) pp. 330–31.
7. Zoya Khaliq Hasan, 'Communalism and Communal Violence in India', *Social Scientist* (New Delhi), vol. 10, 1982, pp. 25–39.
8. Imtiaz Ahmad, 'Political Economy of Communalism in Contemporary India', *Economic and Political Weekly* (Bombay) vol. xix, no. 22–3, pp. 903–6.
9. N. S. Saxena, 'Nature and Origin of Communal Riots in India', *Man and Development* (Chandigarh) Sept. 1982, pp. 25–6.
10. T. N. Madan, 'The Historical Significance of Secularism in India', in S. C. Dube (ed.), *Secularization in Multi-Religious Societies* (New Delhi, 1983) p. 19.
11. For details, see Karan Singh, 'Hindu Renaissance', *Seminar* (New Delhi) (April 1983) pp. 14–18.
12. Lloyd I. Rudolph and Susanne Rudolph, *Economic and Political Weekly* (May 1981) p. 815.
13. Iqbal Narain, 'Cultural Pluralism', p. 916; see also Rajni Kothari, *Politics in India* (New Delhi, 1970) pp. 224–49.
14. Rajni Kothari, 'A Fragmented Nation', *Seminar*, January 1982, p. 24.
15. Bhagwan D. Dua, 'India: A Study in the Pathology of a Federal System', *Journal of Commonwealth and Comparative Politics*, vol. 19, 1981, p. 269.
16. Ibid., p. 269.
17. Myron Weiner, 'Congress Restored – Continuities and Discontinuities in Indian Politics', *Asian Survey*, vol. xxii, no. 4, p. 343; Robert L. Hardgrave Jr., 'India Enters the 1980's', *Current History* (Philadelphia), May 1982, p. 198.
18. Menaka Gandhi's 'Rashtriya Sanjaya Vichar Manch', can be termed as one such party.
19. Weiner, 'Congress Restored', p. 342.
20. Walter, K. Anderson, 'India in 1982: Domestic Challenges and Foreign Policy Successes', *Asian Survey*, vol. xxiii, 1983, p. 128.
21. Weiner, 'Congress Restored', p. 342.
22. S. R. Maheshwari, 'Indian Federal System, Distortions and Corrections', *The Indian Journal of Public Administration* (New Delhi), October–December, 1983, vol. xxix, no. 4.
23. For an excellent analysis, see Dua, 'India: A Study in the Pathology of a Federal System', pp. 257–73.
24. Ibid., p. 271.
25. North-east presently comprises five states, Assam, Meghalaya, Nagaland, Manipur, Tripura and two union territories of Arunachal Pradesh and Mizoram.
26. *Times of India* (New Delhi), 21 September 1984.
27. Ibid., 3 October 1984.
28. P. S. Reddi, 'Genesis of the Assam Movement', in B. L. Abbi (ed.), *North-East Region: Problem and Prospects of Development* (Chandigarh, 1984) p. 259.
29. Estimate of foreign nationals varies from 3 lakhs to 21 lakhs.

30. Maheshwar Neog, 'Assam Agitation against Foreign Nationals' in Abbi, *North-East Region*, p. 281.
31. Fear of government abandonment of the region in the event of a serious national threat continues unabated years after Nehru uttered his anguished cry for sympathy for the people of Assam during the Chinese invasion of 1962.
32. Shanti Swarup, 'Assam: Tensions Arising out of Search for Identity', in Abbi, *North–East Region*, p. 337.
33. D. P. Barooh, 'Silent Civilian Invasion: India's dangers in the Northeast', in Abbi, *North–East Region*, p. 296.
34. The fear of Muslims overrunning the state gets a ready response in view of the fact that Assam is next only to Kashmir in its size of Muslim population.
35. Hardgrave, 'India Enters the 1980's', p. 1175.
36. India Ministry of Home Affairs, *Annual Report*, 1980–81, p. 13.
37. Gen. Mohan Lal, 'Harmony among People of the Northeastern Region and Between it and the Rest of India', in Abbi, *North–East Region*, n. 28, p. 206.
38. Sikh fears were confirmed when the Hindus in the Punjab in a bid to stall the formation of a Sikh majority state, repudiated Punjabi as their mother tongue in the 1961 census.
39. For the full text of the resolution see the *White Paper on the Punjab* (New Delhi, 1984) pp. 79–89.
40. For detailed discussion see K. R. Bombiwal, 'Ethno Nationalism', *Punjab Jounral of Politics*, vol. VII, July–December 1983.
41. *White Paper on the Punjab (1984)*.
42. Anandpur Resolution along with a set of 45 new demands has now been proclaimed as the Charter of Sikh demands.
43. Shorey *et al.*, *The Story of the Punjab* (New Delhi, 1984) p. 183.
44. *White Paper*, no. 39, p. 3.
45. M. V. Kamath *et al.*, *Punjab Story* (New Delhi, 1984) p. 143.
46. *Times of India*, 29 September 1984.
47. Ibid.
48. Ibid.
49. Hardgrave, 'India Enters the 1980's', p. 201.
50. Kothari, 'A Fragmented Nation', *Seminar*, January 1982. p. 26. For a different viewpoint see Bashiruddin Ahmad, 'Threats to the Polity' *Seminar*, August 1983, pp. 34–5.
51. W. H. Morris-Jones, 'India – More Questions Than Answers', *Asian Survey*, August, 1984, p. 813.
52. B. Ahmad, 'Threats to the Polity', p. 35.
53. Morris-Jones, 'India – More Questions Than Answers', p. 813.

3 Pakistan: The Domestic Dimensions of Security

GOWHER RIZVI

One of the crucial problems involved in studying the domestic dimensions of security is to determine what constitutes a threat to national security. In the international arena there is in practice no real distinction between a threat to a state or to the government, for at that level the two are almost interchangeable terms. But when one looks at the domestic sources of insecurity one must make a distinction between the state and the government. It is perfectly legitimate for the opposing political groups to work for the removal of a government and, provided there is an agreed procedure for the transfer of power, it in no way threatens the security of the state. But very often in weak and immature states, an attempt to oust the government is portrayed by the party in power so as to imply a threat to the state.[1] Thus when examining the domestic dimensions of security a major problem is to distinguish between the legitimate opposition to the government and those factors which threaten the continued existence of the state.

An important clue to the problem is to be found in the term *national* security.[2] In other words, any threat to the sovereignty, territory, the ideology and the political institutions of the state which is considered 'national' may be said to constitute a threat to the states' security. This might have provided a useful starting point provided we could define what is 'national' with reasonable certainty. There does not appear to be much ambiguity with regard to the first two, sovereignty and territory; but when we turn to the idea of a state or to its political institutions we are at once on slippery terrain. For an idea or an institution to be the object of security presupposes a national consensus. This is not always so clear cut and governments faced with domestic opposition often use the pretext of a threat to national security or the bogey of foreign involvement to deal with what may be no more than a perfectly legitimate domestic political rivalry. The claim of foreign complicity at once gives the governments the pretext to use force which could not otherwise have been legitimately used against domestic rivals.

The problem of determining what is a 'national interest' is difficult, but not unmanageable, when we are dealing with nation-states or states which have over a period of time evolved into nations. Here the consensus on national interests will be more easily discernible than in those states which have not yet forged a common national outlook. Pakistan which is a multi-nation state, comprising at least four distinct 'nationalities – the Baluchis, the Pathans, the Punjabis and the Sindhis – falls in the latter category. It is extremely unlikely that the four nations will agree on the definition of 'national'. But be that as it may, there are three interlinked components of the state which are generally regarded as the object of national security: the ideology, the territory and the political institutions. The military aspect, even though it remains at the centre of focus of most security analyses, is by no means the only factor, or even the most important one. Political, social, economic and ecological threats deserve equal attention and are likely to provide a more accurate picture of a state's national security dilemmas. In the case of Pakistan the central focus will be on politics because it is our contention that the main threats to Pakistan's domestic security arise from a failure to resolve the political dilemmas of nation building.

In this chapter the main concern is to examine the threats posed to the trinity of the state from the domestic sources. Since external threats are outside the scope of this chapter, the military threats of invasions and blockades are not our concern, except in so far as the military itself is a source of threat to domestic political stability. Thus the Indo–Pakistan rivalry is dealt with in another chapter, and will be touched upon here only because that aspect of security cannot be altogether separated from consideration of domestic security. Similarly the international dimension of the Pakistan–Afghanistan dispute is being considered elsewhere, but because of the religious, national and historical patterns which straddle the state boundaries of the two countries, it will be necessary to deal with that conflict when analysing the Pakhtun and Baluch issues.

The security concerns of a state are defined by the interaction between its vulnerabilities and the threats to it.[3] Not surprisingly, the literatures on security problems often portray 'worst case' scenarios.

Thus studies dealing with Pakistan point to a bleak picture. Her survival is uncertain, her legitimacy is doubted and her ability to defend herself is seen as inadequate. The heterogeneous and centrifugal forces within the country are working to tear the state apart: the presence of a large tribal population, having strong ethnic and linguistic affiliations across the frontiers of both Iran and Afghanistan only adds to the dilemma. Threats to Pakistan are also seen to arise from her geostrategic position, caught up in a global power conflict, and about to be dismantled and partitioned – like a latter-day Poland – out of existence.[4]

Historically speaking, Pakistan, flanked by India in the east and Afghanistan in the west, constituted the north-eastern reaches of the British Indian empire, acting as the guardian of the Khyber Pass against Russian expansion through the buffer zone of Afghanistan. Her relations with her neighbours are far from friendly. With India she seems to be locked in a perennial conflict involving a clash of principles and fundamentally affecting the very basis of her statehood. She has fought three major wars with India and despite being truncated in 1971 has steadfastly refused a subordinate status in the subcontinent. Her frontier with Afghanistan – the Durand line – is disputed and with the Soviet occupation of that country in 1979 and over three million refugees seeking asylum in Pakistan, she is also indirectly in conflict with the Soviet Union. Pakistan's Western allies, with their own axes to grind, have tried to prop up Pakistan, but have proved themselves to be notoriously unreliable. China has been a close ally and has given her much sustenance, but by failing to rally to her aid in 1971 came to be seen as a paper tiger. Although Pakistan is militarily a significant power and she has stood up to India, she is certainly not strong enough, despite American arms supplies, to actually take on the might of the USSR through Afghanistan. The acquisition of nuclear weapons may given her a credible deterrence to ward off her enemies but it would not alter her problem of being overshadowed by a giant neighbour.

But concentrating merely on sources of threats gives a distorted vision, or at best, only a partial picture. The threats to a state can only be studied meaningfully by an understanding of the vulnerabilities of the state as an object of security. Or put another way, a reduction in vulnerability will reduce threat in almost a direct proportion. Conversely, the threats to security will be greatly exaggerated if the state is vulnerable internally.[5] It will therefore be the endeavour of this chapter to focus on threats by assessing them in relation to the existing vulnerabilities of the society.

Security is multidimensional phenomenon and its nature obviously differs from state to state. Pakistan is not a particularly strong state and therefore many of the threats to her security come from within. There are three factors particularly contributing to her domestic insecurity: a narrow and weakly defined purpose of Pakistan in terms of the concept of Islamic states; the absence of consensus on the evolution of national institutions, and the heterogeneous nature of the state.

A coherent idea of the state, either nationalist or ideological is an essential foundation of a strong state for it not only provides the definition, a common bond and the purpose of the state, but also clothes the governing institutions with a legitimacy which gives it authority over its citizens. In the case of Pakistan, formed ostensibly to protect a particular religion, the ideology was never clearly enunciated. The existing works on the origins of Pakistan repeatedly emphasise that Islam constitutes both the reason for its creation and the basis of its existence. While this may be true at a general level, and was a useful tool for organising against the threats of Hindu domination, it was inadequate for organising and welding together a multi-nation state with its bewildering range of ethnic, cultural and linguistic groups. Like the historians of Indian nationalism of an earlier generation, the writers on Pakistan have depicted the struggle for freedom as a monolithic and undifferentiated movement held together by the bonds of Islam. The expression of anti-colonial feelings or a unity forged out of a fear of being swamped by the Hindu culture seems to have been confused as Muslim nationalism.

> 'We are seventy millions and far more homogeneous than any other people in India', wrote the poet Iqbal, 'The Hindus, though ahead of us in all respects, have not yet been able to achieve the kind of unity which is necessary for a nation and which Islam has given you as a free gift.'[6]

And Jinnah's 'two nations' theory declaring that Muslims constituted a separate nation was a logical progression from Iqbal's visions.[7] While it is undeniable that Muslims in India found in Islam an over-arching bond, the potency of which was demonstrated during the successful Pakistan movement, the unity was more apparent than real. Xenophobia and fear of alien culture is a powerful force but a poor material for nation building. Indeed many newly emergent countries have discovered that they have found a state but are still searching for a nation.[8] Jinnah, like Gandhi, had refused to spell out the details of the scheme beyond that Pakistan would be a homeland for the Muslims of India. This had the advantage of keeping the movement from splitting but the crunch came after Independence. The cry of Islam was not in itself enough to reconcile the particularist aspirations of the various ethnic and regional groups.

After Independence the problem of defining the nature of the state could no longer be postponed. What kind of Islamic State would be established? Would it be a theocracy in which the *ulemas* would determine the politics of the state? The poet-philosopher Iqbal who first dreamt up the idea of a separate homeland for the Indian Muslims was emphatic that the Muslim State would not be a theocracy: 'Nor should the Hindus fear that the creation of autonomous Muslim States will mean the introduction of a kind of religious rule in such states.'[9] There were others who were not bothered by social or political philosophies. To them Pakistan simply meant a land of opportunities where they could prosper without competition from the Hindus. A Muslim Pakistan meant a state where land, industries, banks, commerce, civil and military establishments would be under Muslim control. But there were also the radicals who were unlikely to swallow the idea of such a Pakistan. They had been encouraged by Iqbal to believe that 'social democracy in some suitable form and consistent with the legal principles of Islam is not a revolution but a return to the original purity of Islam'.[10] These radicals pressed their socialist demands including centralised economic planning, nationalisation of the key industries, land reform and an equitable distribution of wealth. While they had been emphatic in their support for Pakistan, they were not willing to substitute the Hindu exploitation of the society by that of their co-religionists. G. M. Sayed, onetime president of the Sind Muslim League summed up the feelings:

> Do not forget that Islamic society actually in existence is that in which [the] religious head is an ignorant Mulla, [the] spiritual leader an immoral Pir, [the] political guide a powerful intoxicated feudal lord and whose helpless members are subjected to all the worldly forces of money and influence. If the really important question about the abolition of Jagirderi and Zamindari system crops up or the prohibition of intoxicants become the issue of the day, what would not a rich Jagirdar and an aristocrat member of a sophisticated club do to use his influence, as also that of the Mulla and the Pir, to resist this threat to what is essentially an immoral and un-Islamic cause.[11]

The mere slogan of Islam was clearly not enough. For them Islamic ideology must incorporate social justice and they were not averse to borrowing from Marxist principles. The young Muslim League historian Ishtiaq Husain Qureshi, who later became prominent in Pakistan, was much impressed by the experiments in Russia which he thought had 'a great deal to teach us' and argued that 'all our population will have to be regimented for the purpose of reconstruction which will have to be planned'.[12] It is clear that there was no agreed interpretation of the Islamic ideology and once Pakistan had been won, the soldiers of Islam were less hesitant to break ranks and become vocal in their demand for a clearer definition of the purpose and the idea of the state. The battle was waged both inside and outside the Constituent Assembly.

There were limits to basing the ideology of the state solely on Islam to the exclusion of other factors. The ethnic and cultural identities of the different parts of Pakistan were strong and quite distinctive. They were not necessarily in conflict with the Islamic nationalism of Pakistan, but at the same time they could not be subsumed by Islamic appeals. Even the claim of Pakistan as the homeland for the Muslims of India must have appeared dubious. There were large numbers of non-Muslims living in Pakistan and even larger numbers of Muslims continuing to live in secular India; and when in 1971 the Muslims of Bangladesh asserted their independence and rejected the ties of Islam they dealt another blow to Pakistan's nationalism based on Islam. Even Jinnah, having used Islam to spearhead his campaign, was shrewd enough to realise the problems of making Pakistan a theocracy. His first speech after independence made that clear: 'There are no Muslims or Hindus in Pakistan' he said, 'They are all Pakistanis'.[13] He tried to evolve a common Pakistani nationalism but his early death left the problem far from resolved and the dispute continues to dog politics even today.

Even if the idea of the state is weak, it would not in itself have posed a serious threat to the security of the country. A most cursory glance at multi-nation and federal state structures at once reveals that the decision of different national groups to form a single state is not always based only on ideology or emotional grounds, but often on practical considerations of size, geography, economy of scale and necessities of defence. The early proponents of Pakistan had no doubt visualised a federal scheme with autonomous provinces.[14] Even though the concept and scope of a separate Muslim homeland had never been clearly outlined by any authoritative group or individual, an enquiry into the genesis of Pakistan shows beyond doubt that a federal, or even a confederal, state had been contemplated. Iqbal's oft-quoted 1930 address to the Muslim League where he demanded 'the Punjab, North-West Frontier Province, Sind and Baluchistan amalgamated into a single state' did not include Bengal and he had contemplated 'Muslim states' forming a part of the Indian Federation.[15] Choudhry Rahmat Ali, the Cambridge undergraduate who gave Pakistan its name, had visualised that the provinces constituting Pakistan were to have a separate federation of their own.[16] Dr Syed Abdul Latif's scheme published in 1939 spoke of 'four Muslim cultural zones' in the Indian Federation.[17] Another scheme presented by two professors of the Aligarh Muslim University, Syed Zafarul Hasan and Muhammad Afzal Husain Qadri, divided British India into three independent and sovereign states with the Muslim states of the North-West and Bengal constituting separate states.[18] In 1939 the president of the Punjab Muslim League, in a pamphlet *The Confederacy of India* spoke of an Indian federation which would consist of five states, two of which were to be Muslim.[19] By far the most influential scheme, put forward by Sir Sikander Hyat Khan, the Premier of the Punjab, envisaged a two-tier Federation – a regional and an All Indian Federation.[20] Most important of all the Pakistan resolution of March 1940 itself stated that the 'North-Western and Eastern zones of India should be grouped to constitute "Independent States" in which the constituent units shall be autonomous and sovereign', and that each independent state would have a 'Federal constitution'.[21] The real tragedy of Pakistan and one which makes its territorial integrity so vulnerable, was the failure to evolve a system of governing institutions based on popular consent, in which the different groups in the society could freely participate without the fear of losing their autonomy and ethnic identity. We shall discuss the ethnic question later, but will first focus on the political process in Pakistan. It will be suggested that the insensitive attempts to subordinate the provinces to the centre's direct control and refusal to allow popular participation in policy-making alienated the regional and vernacular élites, who in frustration sought greater autonomy and even secession. To understand this tussle between the centre and the periphery one must again delve into the history of the League and its role in the formative phases of Pakistan.

Although the League had been set up in 1906, it had remained largely dormant until the late 1930s when it launched the struggle for a separate homeland for the Muslims. The movement was a spectacular success. Aided by the special circumstances of the war, Congress intransigence and a helping hand from the British, a student's chimera of 1933 became a reality in 1947.[22] But the strength of the Pakistan movement must not be confused with that of the League. It is true that the League which had polled less than 5 per cent of the Muslim votes in the 1936–7 elections captured over 75 per cent of the Muslim seats in the 1946 elections and even won the referendum in the Congress stronghold at the Frontiers in 1947.[23] But this electoral victory, like the conjuror's trick which deceives the eye by the sleight of hand, was no indication of its organisational strength. The League had never been built-up as a mass party with grass-root organisations. Rather it was organised from top to bottom. Much of its strength accrued from mergers with the regional Muslim parties. As the movement gained momentum, other parties and groups were either swept away or jumped on to the safety of the bandwagon to Pakistan. Many, in fact most, did not share the League's outlook and not surprisingly fell out shortly after the Muslims had homed in.

But far more significantly, and this has not been sufficiently stressed, the League was not really the 'home' party of the areas which actually came to constitute Pakistan. They were, so to speak, playing an away match in Pakistan. In the Punjab the dominant party were the Unionists, who remained hostile to Jinnah until the very end; in the North-Western Frontier Province, Khan Abdul Ghaffar Khan's *Khuda-i-Khidmatgar* were pro-Congress, anti-Pakistan and were pressing for an autonomous Pakhtun state; in Sind, too, the League's control was far from complete.[24] Even in East Bengal, which voted overwhelmingly for Pakistan, Fazlul Huq's Krishak Sramik Proja Party with its commitment to the rural population and agrarian reforms was the real party of the masses and had only agreed to work with the League because the Congress had refused to ally with it in 1937.[25] The League's real stronghold was the United Provinces which became part of India. Many of Pakistan's leaders, including Jinnah, Liquat Ali Khan and Chaudhuri Khaliquzzaman, who dominated Pakistan's politics, were migrants and refugees to Pakistan, without constituencies of their own and lacking links and influence in the provinces. This accident of history proved a major obstacle to the healthy development of the political structure and the centre–province relations. It is the continued failure to resolve these problems which lies at the root of the domestic insecurities that make Pakistan vulnerable to threats of disintegration.

From the very outset therefore a cleavage developed between the popular 'vernacular' leaders from the provinces and the unrepresentative politicians in control of the central government. It was embarrassingly clear, particularly after the routing of the Muslim League in the East Bengal general elections of 1954, that the League's electoral victory in 1946 was a vote for Pakistan and not a mandate for continued rule. In the circumstances, central leaders could not risk a general election to secure a fresh mandate. Instead the rulers at the centre unscrupulously manipulated the emergency powers provided by the Government of India Act of 1935 (under which Pakistan continued to be governed until 1956) and wantonly interfered in provincial affairs and reduced parliamentary democracy to a farce. The provincial cabinets in the Punjab and Sind were dismissed and placed under the Governor's rule; the United Front Government which had won a landslide victory in East Bengal was dismissed within a month after taking office.[26] As their political credibility began to sink, an alliance developed between the unrepresentative members of the Constituent Assembly mainly those whose constituencies remained in India, and the civil bureaucracy, many of whom too were migrants from India. Increasingly the civil servants began to be involved in decision-making often to be the exclusion even of the cabinet members.

Following the assassination of Liquat Ali Khan in 1951, the political stage was dominated by four civil-military bureaucrats – Ghulam Muhammed, Chulan Husain, Major-General Iskander Mirza and General Ayub Khan – who made an utter mockery of parliamentary democracy. Ghulam Mohammed and Iskander Mirza were both experienced bureaucrats and capable administrators. But by temperament and training they lacked sympathy for and understanding of, political processes. Their contempt for parliament and politicians was manifest. They interfered with the workings of Parliament, by-passed the ministers and gave direct orders to civil servants, sacked cabinets without ascertaining whether they actually commanded the confidence of the Assembly, and even appointed their own nominees as prime ministers, securing their support in the Assembly through patronage and coercion.[27] But it must be realised that what gave Mohammed and Mirza this wanton power of interference was the unrepresentative nature of the constituent assemblies. Both the first and second constituent assemblies were elected indirectly on a narrow franchise. Many of the MCAs who had become prominent during the Pakistan movement came from Muslim minority provinces in India and were 'refugees' in Pakistan. They lacked a popular base in Pakistan and when in the early 1950s the popularity of the Muslim League declined, the constituent assembly could no longer be considered as representative of the people. It was not surprising that those politicians with a popular backing would have liked to see a general election to bring the constituent assembly more in line with popular will.[28] But many of the veterans of the Muslim League feared the prospect of facing an electorate and therefore sought safety in an alliance with the bureaucracy who had the backing of the Governor-General. Their common interest also dictated that provincial and vernacular leaders who had mass support be kept out by thwarting any general election.[29]

The parliamentary institutions, which represented a considerable consensus of the élites of both of the wings, were never given a fair run. Although it had taken a long time to frame the Constitution, it was a remarkable essay in parliamentary democracy, allowing for popular participation and with careful safeguards to prevent the Head of State from interfering in politics as had happened in the period up to 1956. It only remained for a general election to be held so that true representatives of the people should take over the control of the government. The civil–military bureaucracy which had got heavily entrenched in decision-making clearly perceived the threats to its position should there be an election. Ayub's military coup of October 1958 was in fact staged to pre-empt the elections.[30] He abrogated the constitution, dismissed the cabinet, banned the political parties, cancelled the elections scheduled for the following spring and usurped the office of President.

Ayub's military intervention was the culmination of a carefully nurtured plan of civil-military bureaucracy to overthrow the democratic institutions which had been in train since 1951.[31] The attempts to do away with both liberal parliamentary democracy and mass participation through adult franchise resulted in a chronic political instability in Pakistan, alienated the representative élites, denuded the government institutions of legitimacy and had led to periodic outbreaks of violence to overthrow the government. Both the Basic Democracy Scheme and the 1962 Constitution were the results of Ayub's *firman*, in no way based on popular consensus. With no agreed procedures for political changes, recourse to mass agitation, non-cooperation and violence has become a common feature. Except for a brief period under Zulfiqar Ali Bhutto when democracy was restored, the military successors of Ayub have tried, with disastrous consequences, to institutionalise their regimes through methods which deny popular participation.[32] The Yahya Khan interregnum led to the secession of Bangladesh; the continual denial of political participation to peoples of all the regions might well provoke further attempts at secession. It would be no exaggeration to say that future political stability, indeed even the integrity of Pakistan will depend on the ability of General Zia-ul Haq to restore government into the hands of democratically elected leaders.

It is perhaps not sufficiently appreciated that the operation of democratic institutions in Pakistan is not a luxury but a necessity for the very existence of the state. Attempts to replace democracy by so-called Islamic or indigenous institutions may be useful rhetoric and a disguise for a naked military rule, but it is a shadowy concept, at best an ingenious device to limit popular participation to a more easily manageable group. Even without challenging the suitability of the so-called Islamic institutions for Pakistan's socio-political and economic conditions, the fact still remains that there has never been a consensus on what is Islamic, let alone what is, to quote Ayub's dubious phrase: 'indigenous to the genius of the people of Pakistan'. All the ethnic groups have their own and very different ideas of what 'indigenous' institutions ought to be. On the other hand, it can be confidently asserted, at least with the benefit of hindsight, that liberal democratic institutions, even if ostensibly imported from abroad, are the only institutions which command consensus and the working of which is familiar to the political élites. The success of India's political institutions is a tribute to Jawaharlal Nehru's perceptiveness to appreciate that secular liberal democratic institutions, even if a product of 'alien genius', was the only one which was acceptable to the Indian political élites.[33] Or to put it somewhat negatively, the secular democratic ideas were less repugnant to the heterogeneous élites of India than would have been an Orriya or a Marathi institution. It is therefore no surprise that most of the Afro-Asian countries which have sought to replace liberal democracy with the so-called native politics have ended in political turmoil or, still worse, in military rule. The list is too long to enumerate.[34] Pakistan is perhaps only one of the most unfortunate examples. But before going any further it might be as well to dispel the notion of democratic institutions being alien to the Asiatics and the Africans. In fact democracy and parliament is no more alien to the Indian culture than is the widespread use of chilli. The commonly known Westminster model was never for export – it can thrive only in the cold and damp climate of the north bank of the Thames – and more importantly it is not a static concept but one which is constantly changing and growing.[35] To that extent it cannot be uprooted and repotted like

Cleopatra's Needle now standing on the banks of the river Thames. The parliamentary institutions operating in Ottawa, Canberra or New Delhi, might have been at one time modelled on the Mother of Parliaments, but in the distant clime and soil, they have become indigenised, drawing succour and strength from their naturalised environment.

In a country like Pakistan, with its heterogeneous peoples and culture, with its unequal economic development, level of education and share in federal services; where resentment against a Punjabi controlled central government and economy is widespread, democratic institutions with full popular participation are essential to create a feeling of commitment to national affairs. The absence of democracy is the root cause of Pakistan's political vulnerability. One of the ways in which the centrifugal forces can be checked is by creating a sense of involvement in national policies and decision-making. But the policies actually pursued in Pakistan seem almost deliberately designed to arouse resentment.

The long, and often cynical debate about the creation of an Islamic polity has helped tottering autocratic regimes to bolster support but failed to arouse a popular response. One does not have to be a Marxist to believe that religion, like hemlock, dulls the senses. But its ability to induce drowsy numbness becomes less as the alchemy of the brain becomes resistant to it. General Zia's attempts to establish a *Nizam-e-mustafa* is simply another variant of the unabashed use of Islam to cloak the illegitimacy of his regime. But even Zia's pious gestures are beginning to wear thin, creating more rifts in the society than they heal. Islam can no doubt provide the basis for a wider international brotherhood to which all Muslims can belong, but it is incapable of supplanting the particularities of its adherents. The Muslims of Pakistan have a twin identity. They are Muslims but they are also Pathans, Baluchis, Punjabis and Sindhis. To ride roughshod over either of the twin identities is courting danger. Both identities are integral to Pakistani nationalism. Neither one or the other has any primacy.

The fears of the eastern wing breaking away had always haunted the minds of the security policy formulators, and until 1971 all eyes were focused on that province. Now, not surprisingly, the gaze has turned to the remnants of Pakistan, reinforced in the knowledge that secession is not merely an academic obsession but something which can actually happen again. But in the true Whiggish tradition whereby the secession of Bangladesh was seen as inevitable, the scholars in turning their attention to the ethnic resurgence in the North Western Frontier Provinces, Baluchistan and Sind, have now discerned numerous strains of centrifugal viruses.[36] The Balkanisation, Polandisation or even Finlandisation of Pakistan are the common but inevitable diagnoses of the learned *Hakims* and *Pandits*. But an important point is forgotten. The first disintegration of Pakistan was the result neither of an organised secessionist movement nor of irredentism but occurred because of the failure of national integration and the alienation of the Bengal counter-élites.[37] The problem of ethnic and linguistic heterogeneity observed in Pakistan is certainly nothing unique: in fact it its a common phenomenon in most Afro–Asian states.

Nevertheless the ethnic problem is an important issue and the nettle must be grasped. The main source of concern comes from the three smaller provinces – Baluchistan, the North-West Frontier Province and Sind. Although each of these regions has its own individual and complex problems, they all share certain common grievances: denial of political participation, undue interference by the federal government, insufficient economic assistance and a resentment against the Punjabis. The Punjabis, in turn, look upon them as traitors and fifth columnists.

The Pathan issue is complex and long-standing.[38] The Durand line which now constitutes the boundary between Afghanistan and Pakistan has been bitterly contested by the Afghans as it bifurcates the fifteen million Pathans almost equally between the two countries. The Pathans who straddle the frontier share strong cultural, religious and linguistic affinities; and their ethnic and tribal loyalties have a primacy over loyalties to the state. They move freely across the Durand line, ignoring the Pakistan government's half-hearted attempts to restrict their movement. Pakistan's authority over the tribal areas is far from complete and is largely dependent on the goodwill of the tribal elders. The relationship is based on the Pathan code – the *Pakhtunwali* – and compliance with it is often reinforced by lavish bribes and gifts to the tribal elders.[39] But Pakistan's efforts to integrate the Pathans is further complicated by the attitude of the Afghan government. They have disputed validity of the treaty of 1893 with the British which established the Durand line, claiming that the treaty defined a sphere of influence rather than an international frontier; and that by the 1921 treaty the British recognised Afghanistan's special interest in the area. Until recently Afghanistan had also refused to accept Pakistan's claim to be a successor state to the

British and thereby denied any treaty rights to Pakistan. They also dispute the validity of the referendum in 1947, which the pro-congress *Kuda-i-Khidmatgars* (who had won a majority in the recent elections) had boycotted because the choice given to them was one of joining India or Pakistan but no option of securing independence.[40] The Pashtunistan issue is also linked to Afghanistan's domestic political calculations. The Pushtu speaking population of Afghanistan, although the dominant group, do not actually constitute a clear majority. The accession of the Pathans from Pakistan would enormously enhance the political standing of the Pushtu speakers.[41] Pakistan, obviously rejects the Afghan claims and suspects that the Afghan support for autonomy of the Pathans is the first step towards full annexation. However, the Afghan government's ability to support Pushtu irredentism is to an extent limited by the fact that Pakistan is a superior military power. Moreover, within Afghanistan the non-Pushtu speakers are opposed to accession of the Pakistani Pathans which threatens to reduce them to tiny minority groups. The situation has been aggravated by the influx of nearly three million Afghan refugees into Pakistan following the Soviet Union's invasion of Afghanistan. But its full implications for Pakistan–Afghanistan relations and for the security of the region will be dealt with in another chapter.[42]

While it is evident that tensions exist in the region, the evidence seems to indicate that the demand of the Pathans is for greater autonomy rather than secession. In fact the Pathans are, compared with the Sindhis and Baluchis, better integrated and have a larger stake in Pakistan's economy, and consequently stand to benefit by remaining within Pakistan.[43] They are well represented in the armed forces, holding about 20 per cent of the jobs while constituting only 13 per cent of the country population; and they virtually control long distance road transport and haulage. The tribal Pathans also carry on a lucrative smuggling trade and operate a country-wide network which extends as far as Karachi. The sturdy Pathans are hard working and loyal workers and through their enterprise have moved into all the areas of Pakistan. In Karachi alone there are over a million Pathans. Negatively speaking, the tribal rivalries among the Pathans are bitter and enduring and thus preclude united action.

The Baluchis, constituting about 5 per cent of the population and occupying about 46 per cent of Pakistan's territory, share certain similarities with the Pathans.[44] They are divided into numerous tribes but share strong cultural and linguistic affinities and straddle across the frontiers both in Afghanistan and Iran. At the same time the Baluchis are deeply divided among themselves, partly de-tribalised, and varying in lifestyle from nomads to city dwellers. In the last thirty years there has been a gradual diaspora of the Baluchis: many have moved to different parts of the country and a substantial number have migrated abroad. Even in important Baluch centres like Quetta, Zho and Laralai the majority of the population is non-Baluch; and Urdu is spoken by many Baluchis. They are economically and educationally backward, are extremely poorly represented in the federal services and particularly in the armed forces, and the region has the appearance of being under military occupation of the distant metropolis. But the area is considered rich in minerals. There is a large gas reserve – the Suigas – and there are believed to be oil deposits. The main resentment of the Baluchis is, in common with the Pathans, an insensitive attempt by the centre to interfere in local affairs and a step-motherly economic treatment. The efforts at so-called modernisation through building of schools and roads has brought mixed blessings. There has been a spontaneous growth of nationalism and many of the leaders have been inspired by revolutionary ideologies in sharp contrast with the religious obscurantism of the military rulers.[45] Baluchi nationalism is largely a reaction to federal policies but at its most expansive it takes the demand for greater Baluchistan, which although no clearly defined, would seem to include territories inhabited by the Baluchis in Pakistan, Iran and Afghanistan.[46] Afghan support for the Baluchis has varied depending on the state of its relations with Pakistan; and the Iraqis have supported the Baluchis to spite the Iranians. The turbulence in the region is further increased by the Soviet penetration in Afghanistan and the turmoil in Iran. Potentially Baluchistan offers the most serious threat to Pakistan integrity but here, as in East Pakistan, the main impetus for secession comes from the Centre's insensitive handling of the genuine demands of the Baluch people. The scope for a foreign power being involved in Baluchistan is considerable.

The Sindhis, who have also been restive recently, have somewhat different grievances. Although they are a separate linguistic group with their own distinctive culture, there is no suggestion of a secessionist or irredentist movement. It is much more a problem of a neglected province and the fear of being supplanted in their own home.[47] Their grievances may be stated under three headings. First, in common with the Pathans and the Baluchis, they resent the Punjabi domination and hold that to be responsible for their under-representation in the central and provincial services, in the armed forces and in commerce. The Sindhis also resent the fact that they have been ousted from some of the best and most fertile lands in the Ghulam Muhammad Barrage area of Sindh which has been used for resettling the ex-servicemen who are mostly Punjabis. The Bhutto government to an extent pacified the Sindhi resentment by distributing the excess land, which became available as a result of lowering the upper ceiling of land ownership, among the native Sindhis. Second, they are also resentful of the large refugee population which has virtually taken over the city of Karachi. The 'new Sindhis', as they are called, are economically and culturally dominant and constitute over three-quarters of the population of Karachi. There have been language riots as Urdu seems to be driving out Sindhi; and there have been clamours for greater Sindhi quotas in jobs, education and business. However, this resentment is somewhat alleviated by the benefits which have been brought to the region by the presence of the 'new Sindhis'.[48] Karachi, which is easily the most advanced and industrially developed city in the whole of Pakistan, has provided greater job opportunities and a higher standard of living. Moreover, the Sindhis have found in the newcomers an able ally in keeping the Punjabis in check. And finally, a new tension has arisen in Sind as a result of Zia's so-called policies of Islamisation.[49] Over 20 per cent of the population of Sind is Shia and hostile to Islamic laws based on the Sunni doctrines. There has been considerable rioting in which even the impartiality of the police and the army became suspect as they did not remain entirely free from sectarian loyalties. Although the Shias are a minority, they are a well-knit community, controlling large portions of trade and industries, banking and insurance, and occupy many senior positions in the armed and civil services and wield an influence significantly beyond their numerical strength. They can also look to Iran for moral support.

The ethnic heterogeneity, racial and tribal tensions or linguistic dissimilarities are not in themselves a sufficient indication of the ability or the desire for secession. What propels a group to break away is the insufficient political and economic integration of the different regions, and where some groups, whether a minority or even a majority, perceive exclusion, discrimination or exploitation by the dominant coterie which controls national decision-making. But our focus on the different groups does inform us of the existence of tensions, grievances and goals of ethnic nationalism. Whether such grievances will be translated into a separatist movement, and whether such a move will be successful, will largely depend on the handling of the situation by the federal government and to a lesser extent on the attitude of the foreign powers. The track record of the central government is one of the almost callous insensitivity and gross neglect. The history of the tussle between the centre (which is largely Punjabi) and the provinces is as old as Pakistan itself.

Shortly after independence a bitter conflict developed in the constituent assembly over the question of regional representation. The Basic Principles Committee, which made recommendations for the draft constitution, laid down that East Bengal with over 55 per cent of the population be accorded 60 seats and another 60 be distributed among the various members from each of the two wings.[50] Although East Bengal was given lesser representation than its numbers merited, they were prepared to accept the recommendations which were in accordance with federal arrangements elsewhere. Curiously the opposition came from the Punjabis on grounds that the Bengalis may get the upper hand. They argued that because West Pakistan was administratively fragmented [consisting of three governor's provinces (Punjab, Sind and North-Western Frontier Province), Bruken Baluchistan, Baluchistan states (Kalat, Las Bela, Malkran and Khairan) and North-West Frontier States (Dir, Amb, Swat and Chitral), the Frontier tribal areas, the princely states of the Khairpur and Bahawalpur and the federal capital of Karachi] whereas East Bengal, being one unit, would come as a united group and by forming an alliance with the smaller groups in West Pakistan, dominate the government.[51] The boot was on the other foot. The self-appointed custodians of Pakistan were after all parochialists when it seemed to threaten their interests. The Punjabis endeavoured to remove any chance of a Bengali supremacy at the centre by undermining the Basic Principles Committee's report. This was not difficult given their control over the civil and military bureaucracy which at this time was increasingly dominatng the decision-making. The upshot was the so-called Muhammad Ali Bogra compromise of October 1953.[52] Each of the five units were given equal representations in the Upper House, and the Lower House was composed on the basis of population. Equal power was accorded to both houses, deadlocks to be resolved by majority votes in joint sessions but the majority would need to include 30 per cent votes from each zone. The Bengali majority was effectively eliminated but still fell short of establishing the Punjabi hegemony. This was achieved by joining the three provices and other areas of West Pakistan into one unit in 1955 despite angry opposition from the provincial assemblies of Sind and the NWFP.[53] The conflict which had at first appeared to affect the relationship between the centre and East Bengal was actually the thin end of the wedge which opened the door for indiscriminate interference in other provinces as well. From their vantage point of the federal government in Karachi (and later Islamabad) the Punjabis could dominate the entire country including East Pakistan; and, after the creation of one unit, from their provincial capital in Lahore they would subordinate Sind, NWFP and Baluchistan. The Punjabi ascendancy was now complete. Greater centralisation which necessarily followed from the creation of one unit, far from allaying regional misgivings, merely increased the tensions. This situation continued unaltered until the one unit was undone under popular protests in 1969–70.

A new era ushered in 1972, when following the defeat, and the loss of Bangladesh, the military was sufficiently discredited to be forced to abdicate power. The restoration of democracy and the formation of popular governments both at the centre and in the provinces were hopeful signs and augured well for the future. Zulfiqar Ali Bhutto was not only an adroit politician, and a sharp intellect, but also the head of a well organised political party. He was the head of the first democratically elected government since 1947 and therefore enjoyed unprecedented legitimacy and popular support. He also had the inestimable advantage of a defeated and a humiliated army, which had badly burnt its fingers and was reconciled to retiring to the barracks – at least for the time being.[54] Moreover, India by declaring a unilateral cease-fire in the war against Pakistan in the western front had clearly indicated that while it had been prepared to act as the midwife in the birth of Bangladesh, it had no such intentions in the western wing. Similarly, that Shah of Iran was keen to ensure the status quo on Pakistan's western frontiers.[55] Thus rid of the military and freed from worries of external intervention, Bhutto could concentrate on the task of pacifying the disgruntled regions. Here Bhutto's effort while moderately successful in Sind, was almost disastrous in Baluchistan and the NWFP. Under the leadership of his cousin, Mumtaz Ali Bhutto, considerable progress was made in recruitment of Sindhis in government services. As the Pakistan Peoples' Party (PPP) controlled both the centre and the province, it was easy for the provincial government to secure large funds from the federal government to meet the demands of the Sindhi nationalists.

But in the NWFP and Baluchistan the situation was different. The PPP had failed to make an electoral mark and the governments were formed in both these provinces by a coalition of the National Awami

Party and Jaminat-i-Ulema Islam (NAP–JUI). The coalitions were stable and the NAP leader Khan Abdul Wali Khan gave no indications of harbouring secessionist intentions. They used the governmental powers and resources to give a booster to the legitimate national, cultural and linguistic aspirations. It seemed that democratic institutions could work with a reasonable harmony even though the parties at the centre and in the provinces were different and somewhat ideologically apart. But the optimism did not last long. Bhutto, who could be both magnanimous and petty at the same time, was unwilling to tolerate opposition from even those who, like himself, were democratically elected. He was not unwilling to stoop to undemocratic methods to undermine the stability of the popularly elected ministries in these two provinces unless they agreed to become junior partners to the PPP.[56] He tried to extend the authority of the central government by appointing the opportunist Khan Abdul Qaiyyum Khan as the federal Home Minister with responsibilities for Tribal Affairs. Qaiyyum, himself a Pathan, was *persona non grata* with the NAP and his relations with Wali Khan were far from amicable. By using his own tribal connections, Qaiyyum sought to de-stabilise the NAP–JUI alliance. In many respects he reactivated the tussle between the centre and the two provinces reminiscent of what had happened prior to 1971. At the same time the constant interference and military actions in Baluchistan had also been causing strain in centre–province relations. When Bhutto dismissed the NAP–JUI alliance in Baluchistan in early 1973, the coalition at Peshawar also resigned in protest. Once again violence broke out in both provinces and was aggravated by the extensive use of the army to pacify the tribal population. When in February 1975 Hayat Mohammed Sherpao, the PPP appointed governor of the NWFP was killed in a bomb blast in Peshawar University, Bhutto launched a massive campaign of repression, and banned the NAP. Once again, in what was to be a repetition of the Agartala conspiracy case in East Pakistan, the NAP leaders including Wali Khan were tried for treason. Wali Khan was accused of viewing Pakistan as consisting of four nations instead of one and working with Afghanistan towards the disintegration of Pakistan.[57] Neither repression nor bribery restored the central government's authority in the two provinces. At the time of Bhutto's overthrow, Baluchistan was in a state of rebellion and a large army contingent was deployed to fight the insurgency, but with little success. Once again the insensitive handling of the ethnic susceptibilities nearly helped to provoke a secessionist movement. The irony was that an elected government at the centre had failed to respect the wishes of the electorate in the provinces. Indeed, the military ruler, Zia, has shown himself much more adept in handling the situation. He has thinned the military presence in the region and has injected large sums of money for development. But the resentment is simmering.

The problem of ethnic nationalism faced by Pakistan is nothing unique. Most Afro–Asian states, as indicated above, are faced with similar experiences. It is extremely unlikely that many, if any, of these ethnic groups will be able to wage a successful war of liberation. Although civil wars and attempted secessionist movements are quite frequent, very few have actually been successful. The case of Bangladesh was unique and is unlikely to be repeated elsewhere. But this is not surprising. Modern states, even the banana republics, ruled by tin-pot dictators, possess vast coercive apparatus and are not unwilling to use repressive measures, including massive genocide, as for instance in Timor, to subdue attempts at secession.

But success at frustrating break-away groups is no measure of the country's domestic security. Preventing a secession is a function of force. States strong on coercion are not necessarily strong on consensus. Indeed, scholars make a fundamental error in looking to the state's ability to hold on to its component parts or the forcible subduing of the population as steps in the direction of building a nation–state. But this is not entirley surprising. Scholars trained in the European tradition perhaps forget that liberal nationalism in Europe in the nineteenth century was a very different phenomenon from the Afro–Asian nationalism of the post-war period. In Europe the nations preceded the state. The Italian and German nations had existed long before their political unification. Unification was merely a *de jure* assertion of national self-determination that had existed *de facto* long since. Afro–Asian nationalism, a rather complicated phenomenon, was to a large extent born of the anti–colonial struggle. The states created as a result of decolonisation were not nation-states based on national self-determination. They inherited the artificial and confused boundaries which had been drawn by the colonial powers, often in ignorance and largely for convenience. However, the independence movements which so successfully expelled the foreign power, and are at pains to remove all vestiges of that hateful past, have rather incongruously held on to the national boundaries defined by that same colonial power, as sacrosanct.

There are bound to be tensions in trying to knot together such disparate regions with their rich diversities as in Pakistan (or India). But the difficulties are compounded by a belief that ethnicity is a problem and impediment to nation-building. It is argued that ethnic nationalism with its fragmented loyalties will gradually wither away in the process of modernisation, thus making room for loyalty to the nation-state. Such assumptions are based on two misconceptions. First, anti-colonial nationalism was based not so much on the Wilsonian principle of national self-determination, as on anti-foreign sentiment which had for the moment papered over the linguistic, cultural and regional sense of identities. The post-colonial states must accord scope for full and unfettered development of the separate cultural aspirations of the minorities.

Second, the climate of nationalism and moderisation has, 'not only encouraged national and supra-national movement, but also new trends towards particularism, such as ethnic and sectarian nationalism'.[5u8] The burgeoning pressure of ethnic nationalism cannot be ignored by any society or political system. But is seems to be an error to assume that there is an inherent conflict between national-states and the various ethnic groups. Nor is there any reason to suppose that the central government must either suppress regional particularities or face disintegration. The possibility of secession, which seems to haunt so many Pakistanis, would appear to be rather remote.

For a secession movement to be successful it would require either the breakdown of the central authority or support from a major power or a superior regional power without a countervailing support for the parent state. After the successful breakaway of East Pakistan, it seems extremely unlikely that India would gain much by supporting (except perhaps as a tit for tat for Pakistan's encouragement to the Sikhs) a splinter group in Pakistan. Moreover, those who live in a glasshouse know the dangers of throwing stones. India's own hands are now full with the Sikhs and encouragement of such centrifugal forces in Pakistan might well resound back across her own border. Besides, with the Soviets in Afghanistan, it would appear to be in the interest of India to have a stable Pakistan as a barrier to keep out a direct involvement in South Asia.

Threats to Pakistan's western frontiers are more complicated. The Soviet presence in Afghanistan has had a double-edged effect for Pakistan's security. On the one hand, the sight of three million Afghans fleeing from the persecution of a 'godless' regime must cause serious misgivings in the minds of the fiercely freedom-loving and fervently Islamic Pathans, about the dangers of breaking away from Pakistan only to come under a Soviet dominated regime with its greater coercive power. And yet, on the other hand, Pakistan could invite the wrath of Soviets, by continuing to harbour Afghan refugees and guerillas thereby making the Soviet task of pacifying the Afghan rebels more difficult.

The Soviet irritation with Pakistan is unlikely to last beyond the present crisis in Afghanistan. In fact it would be in the interest of both Pakistan and the Soviets to reach an agreement, and such an agreement is clearly not impossible. But the problem of ethnic nationalism would appear to be much more enduring and for that reason continues to pose a threat to Pakistan's security. The ethnic problems of language, cultural heritage, autonomy, role of local élites and access to central resources are intricately tied to the question of national consensus and participation in the government. The two cannot be separated. Although the military regime of Zia has shown much greater sense and sensitivity than either Bhutto or any other government hitherto, the fact remains that in the absence of a democratic system both the masses and the élites of the grieving provinces can feel no sense of involvement. The chasm, watered by fear, suspicion and frustration, continues to grow wider and before long the alienation may become complete. It is probable, given the geography of the country and Pakistan's massive armed forces, that the secessionists may not be successful but that will not minimise the inevitable bloodbath or subsequently heal the resentments.

One of the crucial factors on which hinges the security of Pakistan and its ability to shape the future will be its ability to fulfil the economic expectations of the population and generate an equal economic growth in the different parts of the country. There is no denying that at the root of the alienation of the Bengalis was the unequal economic development in the two wings.[59] Moreover, even within the present Pakistan the so-called rapid economic development was achieved at the expense of enormous widening of the gap between the rich and the poor.[60] This was an important factor in the erosion of support which Ayub had built up through his Basic Democracy network. Bhutto had clearly sensed the mood of the people and his spectacular success in Pakistan's two general elections was due to his energetic promise of '*Roti, kapra and ghar*' (bread, clothing and a house), which he picturesquely described as Islamic socialism.[61] Bhutto's five years in office and the spate of rhetorics and reforms have caused a revolution of rising expectations which will have to be fulfilled by any political party aspiring to capture power. In recent years Pakistan's economic problems have to an extent been cushioned by generous aid and loans from the Arab world. The oil boom in the Gulf area has provided opportunities of employment for large numbers of Pakistanis who would otherwise have remained unemployed and formed the nucleus of unrest. Even more significantly the large amount of home remittances amounting to over 2.5 billion dollars in 1984 have eased the pinch of hyper-inflation in the country.[62] But this is a short-term amelioration of the economic problems. The oil bonanza is already levelling off, closing down employment opportunities, and the home remittance is already beginning to decline. In the early 1970s the massive nationalisations of not only banks, insurance and large industries but also of small agro-based and export business was a natural reaction to the unbridled capitalism of the Ayub Khan era. Even though many of the measures were hasty and rushed without adequate preparations, it nevertheless helped to narrow the gap between the haves and the have-nots and to provide a stake in the economy for the upwardly mobile and articulate classes.[63] The military regime which seized power in 1977 took a number of measures to please the business community but did not tamper too much with the public sector. The public sector acts as a sort of safety valve by providing job opportunities to the rising middle-class. The current civilian government, relatively unencumbered by doctrines or electoral promises, seems to be pursuing a sensible policy of freeing the economy from bureaucratic controls. The appointment of Dr Mahbub ul Haq as the Minister of Finances and Planning itself is an indication of things to come. Haq who had risen to fame in the 1960s when he attacked Ayub Khan's development strategy which allowed concentration of wealth in the hands of Pakistan's legendary twenty-two families by impoverishing the masses. Haq is not a socialist. His model for development is South Korea, and he hopes to boost the economy through liberalising trade and giving greater incentives to private entrepreneurs while at the same time seeking to ensure a reasonable standard of living for the masses. Haq knows that this is not a choice but a necessity. Economic progress and social redistribution are crucial for long-term political stability. Pakistan is fortunate in that it has an adequate trained and industrious manpower and important natural resources. The economic prospects are certainly promising.

In economy, as in politics, the establishment of a democratic regime is central to Pakistan's survival. Not only are the resources scarce but there is also a strong competition for a share of the resources between the needs of development and welfare and the demands of the armed forces. Ever since 1947 the armed forces have siphoned off over 60 per cent of the national budget. A poor country like Pakistan cannot afford such wasteful extravagance but given the dominance of the military until recently, the arithmetic of resource allocation could scarcely have been otherwise. However, with the coming to power of a new government the hopes are once again running high.

CONCLUSION

In conclusion it can be said without too much hesitation that despite internal vulnerabilities the chances of Pakistan disintegrating further appear to be remote, unless, of course, an outside power intervenes again. But the chances of that happening despite Indo–Pakistan rivalry and the Soviet occupation of Afghanistan can be discounted for the moment. As has already been argued above the chances of successful secession even in weak states is almost impossible in the last quarter of the twentieth century. The successful breakaway of Bangladesh was in many respects a unique event. The fact that the eastern wing was a thousand miles away from Pakistan and separated by the territory of a hostile power caused serious logistical problems but would not in itself have been sufficient for a successful liberation struggle. Indeed, Pakistan demonstrated its ability to move over one hundred thousand troops of occupation and through a policy of repression and genocide, which is variously estimated to have claimed between one and three million lives, almost succeeded in 'pacifying' the revolt. In the end what enabled Bangladesh to break away in less than nine months of civil war was the decisive intervention of India on the side of the Bengalis and a corresponding failure of Pakistan to secure a similar counter-balance against India. The USA and China made gestures but shied back from actual physical involvement, thus leaving India free to tilt the balance in favour of the Bengalis. The story of Bangladesh is clearly an exception and not likely to be repeated. But the real question to consider is not whether Pakistan will disintegrate, but whether in the long run it is desirable or even worthwhile to hold on to a region perpetually against its will even if it were possible to do so. While it has been argued that Pakistan is politically a weak state, it has considerable coercive power to prevent further secession.

The problems confronting her domestic security are in no way unique. Indeed, very few of the newly emergent Afro–Asian states can be considered as nation-states in the nineteenth century European sense. They are artificial legacies of colonial empires: a ramshackled alliance of heterogeneous groups of people, often linguistically, religiously and racially disparate, who were put together for administrative convenience. These people temporarily united together in their struggle for emancipation from foreign rule but the unity was largely based on xenophobia and therefore superficial. It was therefore not surprising that the unity, forged in the anti-colonial struggle, withered with the disappearance of the foreigner. The common problem in Africa and Asia today is that having found a state, the people are now struggling to forge a nation. Nationalism in the Afro–Asian context is not necessarily based on something concrete like a common history, ancestry, culture, religion or linguistic affinities. It is much more abstract, or even mythical: at its broadest a state of mind which permeates the majority of the people and expresses itself in a desire to live together in a state, the boundaries of which were defined by the colonial rulers. With varying degrees of success the new states are developing institutions, ideologies and socio-economic systems acceptable to the majority of the people, and thereby they are creating a firmer foundation for the state. Herein lies Pakistan's main failure.

In Pakistan, which is a multi-nation state, the process of nation-formation was derailed early first by the failure of the Muslim League to evolve a broad based political organisation along the lines of the Indian National Congress, and subsequently by military intervention in politics. It was believed, falsely as it turned out, that Islam provided the ideology and the over-arching bond for holding the disparate state together. Such an assumption ignored the ties of tribe, language, culture and race, all of which are extremely prominent in Pakistan. What the founders of Pakistan failed to recognise was that Pakistanis had twin identities: they were both a Muslim and a Punjabi, or Bengali, or a Sindhi or a Baluch or a Pathan. To allow primacy of one identity over another was to court disaster. The inability to evolve an acceptable ideology, or to put it differently, the desire to impose the Islamic ideology by ignoring regional sensitivities, was made worse by the failure to evolve an institutional framework in which all the regions could participate equally and not constantly feel exploited or deprived of their rights. Here again the moral of the Bangladesh secession is relevant. It has already been argued that the secession of the eastern wing was neither inevitable nor even inherent in the situation. What propelled the Bengalis to demand autonomy and finally independence was the callous disregard of their cultural sensitivities and a denial of their economic and political aspirations. To an extent some of the problems were to be expected in a country not only so diverse but also with such disparities in regional developments. But such economic disparities and uneven representation in government and military service could have probably been ironed out had Pakistan been able to evolve constitutional governmental institutions based on national consensus and with genuine participation in decision-making by all the groups. To an extent in a democratic polity discrimination or uneven economic development would have been minimised by the need to carry the support of the representatives of the different regions in decision making; and also, if regional disparities did develop, the respresentatives would have themselves to blame rather than any dominant group. The vernacular élites of several regions were alienated by what they perceived to be discrimination against them by the dominant Punjabis. Here lies Pakistan's crucial failure and the root cause of its domestic vulnerabilities.

In Pakistan where the process of national integration was in its early stages, the participation of all the regions would only be ensured in a democratic framework. For example, the absence of Bengalis in the higher echelons of the civil service and their meagre presence in the armed forces or the poor allocation of resources to East Pakistan may have been avoided or, at least, would have been considered less oppressive and exploitative had the Bengali counter-élite been an effective part of the central decision-making. In the absence of democratic government in Pakistan, the exclusion of the Bengalis was almost complete and therefore when the Awami League pressed its demands for autonomy so as to rectify the continued negligence of the eastern wing, it was able to garner the support of the Bengalis from all segments. Without the safety valve of democracy the crisis of Bangladesh became almost inevitable. The danger of a similar situation repeating itself cannot be ruled out. The Punjabis continue to dominate the governance and the economy of Pakistan largely to the exclusion of the smaller provinces, and especially the Baluchis and the Sindhis. The failure to involve all the regions through participation in the government and the legislature will not only aggravate tension but will also invariably sap the vitality of the state and may even open the door for destabilisation and even foreign intervention.

Even though the arguments for the establishment of a fully representative democracy appear to be overwhelming, the actual development in this direction has been abysmally disappointing. Ever since the military coup in 1958 there have been persistent attempts to evolve a political system which would keep the political initiatives in the hands of the civil–military bureacracy to the exclusion of the genuine representatives of the people. Even after the overthrow of Ayub Khan and his Basic Democracy and the loss of East Pakistan, the military rulers of Pakistan would not yet appear to have learnt any lesson. Zia-ul Haq not only overthrew the only democratically elected regime in the history of Pakistan, but has since 1977 been gradually but surely inching his way towards a political system with the same avowed purpose as his military predecessor: ostensibly to civilianise his regime but careful not to part with the custody of real power. After unsuccessfully experimenting with a consultative assembly, Zia was obliged to legitimise his position by allowing a general election. But he pre-empted the powers of the elected representatives by having himself declared as the president through a dubious referendum, amending the constitution (again with doubtful legal validity) so as to make the president all powerful, banning political parties and debarring many of the top ranking political leaders from the election, and finally institutionalising the supremacy of the military in decision-making by the inclusion of the three chiefs of staff in the National Security Council. Even though Ayub Khan's earlier attempts at controlled and limited democracy were a failure and lie at the root of subsequent political upheavals, it would be hasty to consign Zia's constitutional engineering to the dustbin of history. There is no doubt that the constitution has been altered to fit the military's requirement and to that extent it is understandable because a transition period is essential to allow authority to be transferred from the military to elected representatives. But the real test for the durability of Zia's innovations will depend on whether the existing arrangements are a stepping stone towards restoration of full democracy or whether they are yet again blatant device to retain power in the hands of the civil-military bureaucracy and hoodwink the masses.

As yet it is too early even to speculate. What happens next will depend largely on Zia's ability to keep the army from intervening again. The crunch will come when Zia discards his army uniform for civilian garb. If in order to secure the support of the army Zia dilutes the democracy too far he will be repeating Ayub Khan's mistake and no doubt a fate similar to his will also overtake Zia. But on the other hand if the army feels antagonised, the outcome will again be predictable. Despite all possible safeguards which Bhutto had built into the 1973 constitution, the army was not deterred. It overthrew the democratic government at the first opportunity. The future hangs delicately in the balance. The real hope lies in the realisation by both the politicians and the armed forces that only a government responsive to popular sentiments can ensure long term political stability Pakistan can ill-afford another round of turmoil. To say that Pakistan will become a hole in the map would be taking too pessimistic a view. But it is also questionable whether there is any *raison d'être* for a state in which there is a continually raging civil war.

NOTES

1. B. Buzan, *People, States and Fear. The National Security Problem in International Relations* (Brighton, 1983) ch. 2.
2. Ibid., pp. 53–65.
3. Ibid., ch. 3.
4. S. P. Cohen, 'Pakistan' in E. A. Kolodziej and R. E. Harkavy (eds), *Security Policies of Developing Countries* (Lexington, 1982) pp. 93–117.
5. Buzan, *People, States and Fear*, p. 73.
6. Cited in K. B. Sayeed, *The Political System of Pakistan* (Karachi, 1967) p. 33.
7. *The Time and Tide* (London) 19 January 1940.
8. For Pakistan's failure to evolve a composite nationalism see R. Jahan, *Pakistan: Failure in National Integration* (New York, 1972), *passim*.
9. K. B. Sayeed, *Pakistan: The Formative Phase 1897–1948* (London, 1968) p. 104.
10. Cited in Sayeed, *The Political System of Pakistan*, p. 52.
11. G. M. Syed, *Struggle for New Sindi* (Karachi, 1949), p. 216.
12. I. H. Qureshi, *The Future Development of Islamic Polity* (Lahore, 1946) p. 23 cited in Sayeed, *The Political System of Pakistan*, p. 53.
13. *Speeches of Quaid-i-Azam Mohammed Ali Jinnah as Governor General of Pakistan* (Karachi, 1948) p. 10.
14. Sayeed, *The Political System of Pakistan*, ch. 3.
15. Cited in Sayeed, *The Political System of Pakistan*, p. 33.
16. C. Rahmat Ali, *Pakistan the Fatherland of the Pak Nation* (London, 1947).
17. Sir Maurice Gwyer and A. Appadorai (eds), *Speeches and Documents on the Indian Constitution 1921–1947* (London, 1957) vol. II, pp. 444–5.
18. Ibid., pp. 462–5.

19. Cited in R. Coupland, *Indian Politics 1936–42* (Oxford, 1943) pp. 203–4.
20. Ibid., pp. 204–6.
21. L. A. Sherwani (ed.), *Pakistan Resolution to Pakistan 1940–1947* (Karachi, 1969) p. 21.
22. Gowher Rizvi, *Linlithgow and India: British Policy and Political Impasse in India 1936–43* (London, 1978) pp. 89–128.
23. Ayesha, Jalal, *The Sole Spokesman. Jinnah, the Muslim League and the Demand for Pakistan* (Cambridge, 1975) pp. 134–8 and 171–3.
24. Ibid., pp. 168–71.
25. A. S. M. Abdur Rab, *A. K. Fazlul Huq, Life and Achievements* (Barisal, 1966) p. 89; see also Jumaira Momen, *Muslim Politics in Bengal: A Study of Krishak Proja Pary and the Elections of 1937* (Dacca, 1972), *passim*.
26. Sayeed, *The Political System of Pakistan*, p. 73.
27. For an excellent analysis, see L. Ziring, *Pakistan, The Enigma of Political Development* (Folkestone, 1980) chs 3 and 4.
28. G. W. Choudhury, *Democracy in Pakistan* (Dacca, 1963) ch. 5.
29. Ibid.
30. Gowher Rizvi, 'Riding the Tiger: Institutionalising the Military Regimes in Pakistan and Bangladesh', in C. Clapham and G. Philip (eds), *The Political Dilemmas of Military Regimes* (London, 1985) p. 204.
31. Ibid., pp. 202–4.
32. Ibid., pp. 201–36.
33. S. Gopal, *Jawaharlal Nehru, A Biography* (London, 1979) vol. II, pp. 303–6.
34. See C. Clapham and G. Philip (eds), *The Political Dilemmas of Military Regimes*: W. F. Gutteridge, *Military Regimes in Africa* (London, 1975).
35. A. F. Madden, "Not for Export": The Westminster Model of Government and British Colonial Practice, *Journal of Imperial and Commonwealth History*, vol. VIII, no. 1 (October 1979) pp. 10–29.
36. S. M. Burke, 'South Asia Yesterday, Today and Tomorrow', in H. Wriggins (ed.), *Pakistan in Transition* (Islamabad, 1975) p. 240.
37. Jahan, *Pakistan: The Failure in National Integration*, *passim*.
38. K. B. Sayeed, 'Pathan Regionalism', *South Atlantic Quarterly*, vol. LXIII, no. 4 (Autumn 1964) pp. 478–506.
39. For an excellent study of the Pathans see O. Caroe, *The Pathans* (London, 1958); A. S. Ahmed, *Social and Economic Change in Tribal Areas* (Karachi, 1977).
40. Jalal, *The Sole Spokesman*, p. 290, fn 162.
41. Z. Khalilzad, *Security in South Asia 1. The Security of Southwest Asia* (Aldershot, 1984) p. 141.
42. See below, Chapter 6.
43. S. P. Cohen, *The Pakistan Army* (Berkeley, 1984) pp. 42–5; F. Ahmed, *Focus on Baluchistan and the Pashtun Question* (Lahore, 1975) p. 107.
44. Khalilzad, *The Security of Southwest Asia*, pp. 147–55.
45. T. Ali, *Can Pakistan Survive? The Death of a State* (Harmondsworth, 1983), see appendix 2: Interview with Murad Khan, pp. 200–9; Harrison, *In Afghanistan's Shadow: Baluch Nationalism and Soviet Temptations* (New York, 1981).
46. Khalilzad, *The Security of Southwest Asia*, p. 147.

47. Ziring, *Pakistan*, pp. 142–8.
48. Khalilzad, *The Security of Southwest Asia*, pp. 65–6.
49. S. A. Kochanek, *Interest Groups and Development. Business and Politics in Pakistan* (Delhi, 1983) p. 316.
50. *Report of the Basic Principles Committee* (Karachi, 1952); for a detailed analysis of regional representation, see K. Callard, *Pakistan. A Political Study* (London, 1958) pp. 155–93.
51. Sayeed, *The Political System of Pakistan*, p. 68.
52. Ibid., p. 72.
53. Callard, *Pakistan*, pp. 183–93.
54. Rizvi, 'Riding the Tiger', p. 218.
55. K. Prasad, 'Pakistan-Iran Relations', in S. Chopra (ed.), *Perspectives on Pakistan's Foreign Policy* (Amritsar, 1983) pp. 340–41.
56. Ali, *Can Pakistan Survive*, ch. 5.
57. Ziring, *Pakistan*, pp. 155–9.
58. Khalilzad, *The Security of Southwest Asia*, p. 60.
59. K. B. Griffin and A. R. Khan (eds), *Growth and Inequality in Pakistan* (London, 1972) pp. 1–21.
60. Kochanek, *Interest Groups and Development*, p. 183.
61. Ali, *Can Pakistan Survive*, p. 84.
62. The *Financial Times*, 29 November 1984.
63. Kochanek, *Interest Groups and Development*, p. 324.

Part III
The Regional Component of the Security Problem

From the chapters in the preceding part it has become clear that both India and Pakistan have substantial domestic security problems. Separatist movements threaten national cohesion, governing institutions suffer serious flaws, and in Pakistan, prolonged and repeated military rule has meant that the government has only shallow roots. But both countries have had some success in establishing an idea of the state, both draw cohesion from national economies, and both have political institutions and coercive state apparatus that are powerful enough to maintain the state against internally generated centrifugal forces. Although the two states look durable in their own right, however, it is clear that their domestic political instabilities share some intimate linkages across international boundaries, and that these linkages feed upward to have a major effect on relations between states at the regional level.

In this part, our focus is on that regional level: the dynamics of the local security complex in South Asia. Our attention is therefore on the interactions between states, particularly the ways in which they threaten or enhance each other's security. We have two tasks in this context, first to understand the rivalry between India and Pakistan that defines the complex, and second, to see how the smaller states fit into this pattern.

In analysing the rivalry between India and Pakistan, we examine its causes, the depth of the hostility, and the trends favouring one side or the other. We try to determine whether the rivalry is based primarily on internal factors within both countries, or on a set of issues between them, or if it is a political legacy of partition, with the domestic stability of each being threatened by the very existence of the other. To understand fully the complexity of the rivalry, we raise a number of other questions. An enquiry is made to find out if the rivalry is a reflection of the government's need in each country to maintain internal cohesion by appealing to a foreign threat, and whether the almost continuous military rule in Pakistan exacerbates the tensions already existing between the two countries. Two further issues are raised: first, the role of transnational ethnic, religious, linguistic and cultural ties underlying the conflict; and second, the extent to which the rivalry reflects external issues in the form of territorial disputes, or military balance of power considerations.

In relation to the smaller states, the principal concern is to find out how they fit into the pattern created by India and Pakistan. We enquire into what difference their alignments make to the rivalry between the two larger states, and whether the principal role of the smaller states within the complex is the way in which each relates only to one of the larger states. Looking ahead to the discussions of higher levels in Parts IV and V, we examine the way in which the smaller states provide an avenue for penetration by external powers into the region, and assess what impact this has on the local complex. Because of the close linkages between small states and the domestic security issues of the two large states (especially India, less so with Pakistan), it is necessary to establish the extent to which both the foreign and domestic policies of the small states constitute a problem for the larger ones.

4 The Rivalry Between India and Pakistan

GOWHER RIZVI

The rivalry between India and Pakistan is to a large extent embedded in the structure of the relationship between the two states. While no conflict is ever inevitable, the handling of the problems by the leaders of both the countries, many of whom view Indo–Pakistan relations as a zero-sum game where the gain of one is seen as the loss of the other, has ensured that the rivalry persists, and that it has become deeply ingrained in the politics of the two neighbours. The rivalry is too deep-rooted either to disappear easily or to be capable of a rational solution.

BACKGROUND

While the simple explanation that the Hindu–Muslim communal conflict is the legacy of the colonial strategy of divide and rule ought to be buried and forgotten,[1] it is nevertheless true that the coming of European rule, albeit in an unpremeditated way, contributed to the disharmony between the communities. The East India Company, the predecessors of the British rulers, first entered India through the three ports of Calcutta, Madras and Bombay. The people with whom the British came into contact in the course of their trade, and consequently the Indians who benefited most from the early contact with the Europeans, were mainly non-Muslims.[2] While the explanation for the subsequent Muslim educational and political backwardness is complex and varies from region to region, it would be no exaggeration to suggest that the Hindus of the three Presidency towns of Calcutta, Madras and Bombay had several generations head start in Western education. By the time the Indian National Congress was founded in 1885 and pressing forward its demands for a share in the governance, the Muslims were barely beginning to come out of their seclusion to acquire Western education. Thus from the very start the Muslims trailed behind the Hindus and therefore, and not surprisingly, the two groups were out of step and their demands were at variance.[3] The dissonance was exaggerated by yet another historical accident. In many parts of India the class divide also coincided with the communal line and thereby often distorted what might have been ordinary class or sectional conflict of interest into communal rivalry.[4] The Indian National Congress (INC) had too diverse a following ever to evolve a coherent social and economic programme, and therefore ducked the communal issue – tumbling from the frying-pan into the fire but refusing to articulate a programme which would appeal to the masses of all communities alike.[5] It therefore required little effort on the part of the Muslim leaders to walk off with the Muslim flock when Hindu/Muslim politics began to polarise.

The polarisation occurred not so much because of irreconcilable incongruities or any deep-seated animosities but merely because of irritants and apprehensions which the minority community suffered. It was largely psychological. Minority groups in all societies perceive a threat and are anxious to preserve their separate identity and culture. Whether such perceptions by the Indian Muslims are real or imagined is immaterial. Fears can be genuine without being well based. What is important is that the Muslims harboured such anxieties, and the majority community, or their leaders, did little to disarm them. The Congress argued with considerable logic that reglgion ought not to matter in politics and claimed that there was little to distinguish an Indian Muslim from a Hindu.[6] This of course was largely true. The bulk of the Indian Muslims were converts who had foresaken the religion of their ancestors to escape the tyrannies of caste inequities. They professed Islam but in their habits, customs, mode of dress, diet, language, there was little to separate them from their Hindu neighbours. Even though the mosque and the temple stood apart, the members of the two communities shared common festivals, evolved a common mythology where Ali and Ram, Sita and Fatima are cast in similar roles; and at a popular level they have the same superstitions and worship common deities and saints. But these anthropological observations should not be pursued too far.[7] While the two communities have coexisted peacefully for centuries, and there were no reasons for doing otherwise, they *never* integrated. Certain fundamental differences remained. There were few intermarriages – surely one of the best ways to integrate; to most orthodox Hindus and Muslims, the one polluted the other; after a thousand years in India the Muslims still bear Arabic or Persian, rather than Indian names; and there is a vague but deeply held belief among the Muslims of the sub-continent that West Asia is the repository of spiritual and cultural values.[8] To say this is not to deny that the Muslims were Indians, but merely to argue that like many other communities they had a twin identity: they were Indians and they were Muslims. To assume the primacy of one identity over another was courting friction. In a sub-continent so diverse as India, the ethnic, religious and linguistic particularities have to be accommodated and not subsumed in a larger over-arching nationalism. Muslim nationalism was not inimical to secular nationalism, but merely that it was anxious not to be steam-rolled by the majority community. It therefore sought safeguards either through weightage and separate electorates in the provinces where they were in a minority; or by seeking greater autonomy to manage their own affairs in provinces where they were in a majority and therefore capable of controlling governments.[9] It was the failure of the Congress to respect these susceptibilities which prompted Muslim separatism. Similarly the main failing of Pakistan's nationalism has been its determination to forge a nationalism by appealing to Islamic sentiments to the exclusion of regional particularism.[10] In 1971 Pakistan lost Bangladesh because it forgot the Muslims in East Pakistan were also Bengalis. The message still does not seem to have filtered through: the central governments, both in India and Pakistan continue to be insensitive to the ethnic groups. The demand for autonomy in itself does not lead to secession. It is the failure to respond to such a demand that provokes secession.

The founders of Pakistan, unlike the present day rulers, had not contemplated an Islamic state. What they were seeking was a system of government under which the Muslims could develop their own culture, and their way of life without the fear of being swamped.[11] The introduction of dyarchy in 1920, and the modest freedom it allowed to the provinces, brought home to the Muslims the realisation that while they were a minority overall in India, they were clear majorities in the Punjab and Bengal. The logical implication of this discovery, particularly as the British devolved more powers to the Indians, was that if the Muslims could hold their own politically in Muslim majority areas they need have little fear of being over-run. Henceforth, the Muslims of Bengal and the Punjab were no longer concerned with 'safeguards' or 'weightage', but now demanded greater autonomy for the provinces, and the creation of two new provinces of Sind and the North-Western Frontier Province which would have a Muslim majority.[12] The Congress, obsessed with the fears of centrifugal tendencies could not appreciate the strength of Muslim sentiments. Its response was the Nehru Report of 1929 which rejected both separate electorate and autonomous provinces by calling for a unitary state.[13] In retrospect, it would appear that if the Congress had deliberately tried to drive the Muslims away from the Congress they could not have done better! At a stroke they alienated the Muslims of the minority provinces by refusing separate electorates and the Muslims of the majority areas by insisting on an overbearing central government. Although the tortuous path to Pakistan lay far in the future and was the result of many complex and interacting dynamics, in its essence the demand for a separate homeland for the Muslims was the result of the failure of the Congress to accommodate the Muslim demands for autonomous provinces.

By the time India was partitioned in 1947 considerable venom had been injected into Hindu–Muslim relations. It appeared that only a division of India would restore sanity and a semblance of decency. Pakistan, born in a hostile environment, was designed to prevent the Hindus and Muslims from getting at each other's throats. But the problems were not solved. The creation of Pakistan merely elevated the inter-communal fight into an inter-state conflict. Moreover, the very *raison d'être* of Pakistan as a separate homeland for the Muslims seemed doubtful. Over one-third of the Muslims remained behind in India, casting serious doubts on the validity of Mohammed Ali Jinnah's 'Two-Nations theory'. Jinnah had created a state but the search for a nation and a national identity had only begun.

Pakistan's quest for a separate identity acquired an urgency because many leaders in India appeared to have only accepted the partition, either as Nehru did, to avoid further brutalities and savageries which were being perpetrated by the frenzied members of both communities; or as some others did who saw Pakistan as a temporary aberration, the acceptance of which was necessary to rid India of the British. Many doubted the viability of Pakistan and expected that once the passion subsided, Pakistan would come back into the fold of India.[14] Expressions of such sentiment, privately or publicly, only aroused the fears of Pakistan's leaders which the partition was designed to still. In this hostile environment, as was to be expected, Pakistan played up its Islamic sentiments. This had a dual advantage: it provided a convenient bond for Pakistan's five diverse provinces but also posed a threat to India's professed goal of secularism. The fear of being reamalgamated into India provided the other strand of Pakistan's nationalism.

Despite the holocaust that accompanied the partition, it is possible the two neighbours could have evolved a *modus vivendi*, learnt to live in peace, and even built bridges through collaboration in overcoming problems common to both. The events, however, turned out to be different. The two countries, shortly after independence, went to war over the princely state of Kashmir. Indeed, the dispute over Kashmir reactivated all the issues and the traumas which the partition was intended to stop, and made normal relations between the countries well-nigh impossible Kashmir institutionalised in a microcosm all the historical irritations between India and Pakistan and has continued to defy all rational solutions. The origins of the conflict have many ramifications and are complex.[15] Each side has not only a different story but also convincing arguments to support its claims. Since Kashmir lies at the core of the conflict, it would be just as well to examine the tangled web.

THE KASHMIR DISPUTE

The princely states of Jammu and Kashmir were under the rule of Maharajah Hari Singh. While the ruler was a Hindu, the population of Kashmir was predominantly Muslim. According to the 1941 census of the total population of 4 021 616, Muslims accounted for 3 100 000 and the Hindus 809 000 approximately. When the British transferred power to India and Pakistan in 1947 there were certain ambiguities over the future status of the princely states. The Cabinet Mission Plan of 16 May 1946 had merely stated: 'Paramountcy can neither be retained by the British Crown nor transferred to the new Government'.[16] Lord Mountbatten, was so occupied in trying to persuade the Congress and the League to accept the partition plan that he did not turn to the princely problems until very late in the day.[17] The Indian Independence Act of 1947 had, in theory, left the states legally independent when 'the sovereignty of His Majesty over the Indian States lapses'.[18] But in practice such independence was ruled out when the Secretary of State for India, Lord Listowell declared: 'We do not, of course, propose to recognize any states as separate international entitites.'[19] This meant in reality that the states would have to seek accession to either one of the two Dominions in accordance with the broad principles of the partition itself: Muslim majority states located in territories contiguous to Pakistan would accede to Pakistan and the rest would go to India. In the event five hundred odd states were integrated without any major difficulty. In three cases, Junagadh, Jodhpur and Hyderabad where the rulers tried either to remain independent or accede to Pakistan in breach of the principle of partition, India used force to set right the anomalies. When the Muslim ruler of Junagadh, with nearly 80 per cent Hindu population, applied to Pakistan for accession and was accepted, Mountbatten tried to persuade the ruler that such an accession was 'in utter violation of the principles on which partition of India was agreed upon and effected', and advised that 'normally geographical situation and communal interests and so forth will be the factors to be considered'.[20] The Maharajah of Jodhpur, was similarly admonished by Mountbatten because the subjects of his state being predominantly Hindu, accession to Pakistan 'would surely be in conflict with the principle underlying the partition of India on the basis of Muslim and non-Muslim majority areas'.[21]

In the circumstances it would have appeared that Kashmir too would easily be disposed. Over 75 per cent of the population was Muslim and the state was adjacent to Pakistan and irrespective of the wish of the ruler, the state would be integrated with Pakistan. Maharajah Hari Singh at first delayed and concluded a standstill agreement which Pakistan accepted but India did not reply. The experience of other states had shown that the option of independence even if available in theory was ruled out in practice. But Kashmir, with common frontiers both with India and Pakistan, had some leverage which the Dogra ruler intended to exploit. He had strong reasons for not acceding unconditionally to either India or Pakistan. His regime was extremely unpopular with both the main political parties of Kashmir, the pro-Pakistani Muslim Conference and the pro-Indian National Conference. While as a Hindu he was unlikely to find favour with Pakistan but at the same time, with over three-quarters of his population Muslim, an accession with India would be popularly unacceptable. And finally, the National Conference which was aligned with the Indian National Congress was secular and socialist – with very little sympathy for the princely ruler and therefore the prospect of joining India was not particularly attractive for the Maharajah. In the circumstances Hari Singh dithered. But between August and September 1947 the situation deteriorated rapidly when the Muslim subjects of the Maharajah rose in an open revolt and were soon joined by fellow Muslim tribesmen from the North-Western Frontier Provinces. The overthrow of the Maharajah seemed almost imminent.[22] He fled from Srinagar and in desperation (and prompted by V. P. Menon, the Government of India's Secretary for the Princely States), agreed to accede to India on 26 October 1947. As soon as India accepted the accession it despatched airborne troops and reinforcements to Srinagar and after some heavy fighting was able to quell the rebellion and drive out of Srinagar the invasion of the Pakistani tribesmen. By the time the cease-fire was agreed, India was in control of over two-thirds of Jammu and Kashmir, only a third remained in Pakistan's hands.

Mountbatten, in accepting the accession, made a fundamental error of judgement and left behind the bitterest legacy of British rule for South Asia. He was right when he said 'that in the case of any state where the issue of accession has been the subject of dispute, the question of accession should be decided in accordance with the wishes of the people'.[23] However, in view of the three recent accessions against the expressed will of the rulers and Mountbatten's own advice to them quoted above, the need for holding a plebiscite in Kashmir would seem to be dubious. Moreover, it made little sense for India to try to cling to Kashmir so soon after accepting the principle of partition and indeed even pushing it to its logical conclusion by partitioning the Punjab and Bengal. Kashmir not only had an overwhelming Muslim majority, it was also contiguous to Pakistan, with its rivers and natural lines of communication linking with Pakistan; historically, culturally and economically it was closer to Pakistan than to India. Against all logic and, probably against its own self-interest, India decided to hold on to Kashmir. Her motives are not clear. One can only conjecture. It is possible that Mountbatten, whose *amour propre* had been bruised by Jinnah's refusal to let him be the joint Governor-General of India and Pakistan, might have wished to teach Jinnah a lesson. But how does one explain the acceptance of a plebiscite in sure knowledge that if one were held, the decision would invariably go against India? Probably, the Indian leaders were convinced that either the 'moth-eaten' Pakistan would not survive or that it would be so weak that it would succumb to India's dictates.

When Jinnah met Mountbatten to hammer out the arrangements for a plebiscite which both Mountbatten and Nehru had promised,

India began to procrastinate. Nehru had 'fallen ill' just before the meeting and the Indian Cabinet, fearful of losing the plebiscite, were not keen on negotiating. In the circumstances, Mountbatten dragged his feet as he could 'not venture to take on the responsibility for working out the exact details of the arrangement'.[24] In January 1948, India, rather inexplicably, in view of its subsequent conduct, called in the United Nations but rejected its proposals for arbitration. The UN made several attempts, as did the Prime Ministers of the two countries, but the gulf was too wide to be bridged.[25] The prospects of a settlement receded.

Notwithstanding its promise of a plebiscite, India quietly took measures to assimilate the state of Jammu and Kashmir into the Indian Union. Elections for a Constituent Assembly were held in the Indian part of Kashmir in 1951. As the pro-Pakistani Muslim League boycotted the polls, the pro-Indian Sheikh Abdullah's National Conference Party won all the seats.[26] This was interpreted as a vote for India and the State was given a special status in the Indian Union. But the internal situation was far from satisfactory. The Maharajah Hari Singh was deposed in August 1953. Sheikh Abdullah, the Chief Minister since 1948, was arrested as he pressed his demands for a 'full autonomous' state guaranteed by both Pakistan and India.[27] In February 1954, the Constituent Assembly of Kashmir accepted the recommendation of the Basic Principles Committee to remain acceded to India. And finally, in November 1956, the Constituent Assembly ratified the Maharajah's instrument of accession of 1947 and the state became 'an integral part of India'.[28]

Pakistan was far too weak to make India abide by its promise to hold a plebiscite. As the UN had proved to be without teeth, Pakistan tried to strengthen her position by turning to Britain and the Islamic states. The former was unwilling to antagonise India for the doubtful advantage of cultivating Pakistan; and the latter were themselves weak and in no position to aid Pakistan. Moreover Egypt and Indonesia were India's partners in the non-aligned movement and showed little sympathy for Pakistan.[29]

More crucially Pakistan was also going through a period of domestic crises since the assassination of her first Prime Minister Liquat Ali Khan in 1951. The civil and military bureaucracy together with the unrepresentative politicians jockeyed for power at the centre by excluding the populist leaders of the provinces.[30] It was with this back-drop in Pakistan's politics that the ambitious and scheming Commander-in-Chief, General Ayub Khan, renewed his offer to the

United States of Pakistan's willingness 'to provide the cordon sanitaire around the Soviet Union' and entered into a Mutual Assistance Agreement in May 1954.[31] His motives in seeking military assistance had little to do with Kashmir and still less with fighting communism but were largely to strengthen and modernise the armed forces so that he could wrest political control.[32] But in Pakistan the alliance with the USA was sold to the public in terms of strengthening the country's defences against India. The fact that the USA had stipulated that the arms could not be used against a non-Communist power were glossed over, and provided India with the pretext to renege on its promise of the plebiscite in Kashmir.[33] India probably feared that with the acquisition of US weapons Pakistan would be less malleable, and therefore that it should be kept out of Kashmir.

The stalemate in Kashmir remained and continued to be the major source of animosity between the countries. Yet solutions which were practicable, which would have safeguarded the interests of the poeple of Jammu and Kashmir, which would not have sacrificed India's strategic interests or even her secular principles, and for that matter which would have been acceptable to Pakistan, were not entirely wanting.[34] There were proposals for independent Kashmir; for a condominium; integration of Jammu (with its Hindu majority) and strategically important Ladakh with India; 'Azad' Kashmir and Baltistan with Pakistan with either an autonomous Kashmir valley or a limited plebiscite to ascertain the wishes of the local populace; and finally, of course, the original offer of a plebiscite for the entire region under international auspices. While all these were viable solutions and would have given both countries an opportunity to extricate themselves without loss of face, they were turned down. The real problem was that Kashmir was not merely a territorial dispute but was deeply intertwined in the domestic politics and ideologies of India and Pakistan.

Pakistan's domestic affairs were in shambles. At first the stalwarts of the Pakistan movement who were mainly migrants to Pakistan clung to power at the centre by avoiding elections and excluding the vernacular populist leaders from the provinces. So precarious was their position that no general elections were allowed in Pakistan for nearly a quarter of a century and hence no truly popular government held office until 1972.[35] After 1958 (save for the brief Bhutto regime of 1972–7), the military rulers were too preoccupied with their own survival and lack of legitimacy to take a bold initiative on an emotive issue like Kashmir. Instead, they fanned the Kashmir flame to distract popular attention away from the real issues and problems confronting the country. Moreover the continuing turmoil in Pakistan, together with the absence of democracy, did little to enthuse the Kashmiris to take up arms against India and in Pakistan's favour. Nevertheless, the domestic constraints in Pakistan are such that the rulers would rather risk the economic ruin of the country than give up their claim to Kashmir. While poverty and hunger stalks the country, over 60 per cent of the national budget is allocated for defence. Kashmir has not been liberated but in the process the army has become politically dominant and democracy has been largely buried.

India, whose claims to Kashmir are at best dubious, seems to have still fewer reasons to cling to Kashmir. In fact, in occupying Kashmir India has created a rod for its own back and the full implications of the damage, political and economic, for India will be difficult to measure. Her arguments for holding on to Kashmir are hollow, fallacious, shifting and confused. India had not only accepted the principle of partition in 1947 but also pressed for a logical extension of that principle by dividing the Punjab and Bengal. And yet a year later her insistence that her ideology of secularism is at stake if she relinquished the possession of the predominantly Muslim territory is difficult to comprehend. Nor is her claim to Kashmir's accession by a treaty any more valid. In the first place Mountbatten had blundered into accepting the request for accession when he had himself presided over the rejection of similar accession by the State of Junagadh; and even if such accession were valid, it was conditional upon a plebiscite which never took place. The high moral principle of treaty obligations is of course a sham and sheer hypocrisy; and an admission that India's claim would not be sustained in an impartial referendum. And finally to claim that US military aid altered the geostrategic situation of the sub-continent and to use that as an excuse for India to renege on its international commitment is a sleight of hand too clumsy to deceive the eyes. But these are not the only problems. Far more serious is the impact of Kashmir on India itself. Internationally, the constant bickering between the two neighbours has denied India the role which Nehru had aspired to play and the status that was due to this great Asian state. Economically India has had to pump scarce resources beyond all reasonable proportion into defence to fight Pakistan. In internal politics the existence of Kashmir in the Indian Union with its special status has further exacerbated the cohesion of this heterogeneous state and has often strained the centre-state relations almost to breaking point. With the passage of time public opinion in India has grown to look upon Kashmir as a part of India and no Indian leader can contemplate a compromise without risking his or her life and career.[36] And finally, her relations with the smaller neighbours too have suffered because of her conflict with Pakistan. The smaller states, not surprisingly, see India as an imperialist intent on establishing its hegemony and consequently have sought safety through alliance with outside powers.

Both countries have got themselves into such a position that there was little room for manœuvre. India is in physical possession of two-thirds of Kashmir and while laying claim to the whole, is content to maintain the status quo. But for Pakistan giving up its claims would be tantamount to abjuring its *raison d'être* as the homeland of the Muslims. As diplomacy, mediation, international pressure and alliance had failed; and also because domestic constraints precluded any compromise, Pakistan's only recourse to end the status quo was through war.

When Ayub Khan staged a coup in October 1958 he had achieved his political objective. A shrewd manipulator he realised at once that there was little to be gained through a posture of confrontation with India and addressed himself to easing the Kashmir issue. Although he was the architect of the US military aid to Pakistan and as the commander-in-chief had done much to strengthen Pakistan's military position, he was anxious to avoid a military confrontation with India.[37] His desire for a peaceful settlement was the result of a pragmatic appraisal of the situation. He recognised that despite the massive military build-up in Pakistan, it would not be easy to defeat India. At best the war would end in a stalemate and it might therefore destroy his political credibility. Moreover, his ability to survive in power would depend upon his ability to reduce the burden of the ruinous defence expenditure which had been crippling the country's economy. His was also the approach of a simple-minded soldier.[38] He reasoned that Pakistan's claim to Kashmir was so impeccable and the failure to resolve it would be so disastrous for both India and Pakistan, that Nehru could hardly refuse to compromise. Ayub had underestimated the Indian national commitment or for that matter Nehru's emotional attachment to Kashmir. To Ayub, Kashmir was the core of the hostility. Remove the Kashmir dispute and all would be well. Nehru, however, viewed it differently. To him the conflict in Kashmir was a symptom of a much deeper malaise and therefore he considered it futile to attempt to solve Kashmir without addressing the fundamental issues of conflict.[39] What these issues were was not elaborated by Nehru. But he was in a difficult situation. A high principled statesman who constantly lectured others on international morality, the architect of non-alignment and the *Panch Sheel*, Nehru was batting on an extremely sticky wicket, and called for some quick foot-work. He had repeatedly promised a plebiscite and yet during the decade much water had flowed down the Chenab, and Kashmir had come to be regarded as an integral part of India. He therefore stone-walled the negotiations by demanding a complete rather than piecemeal settlement of the disputes between India and Pakistan. Nehru wanted a draw and was batting to play out the time; Ayub was bowling to force the pace in order to secure a victory. To drop the cricketing metaphor, Nehru wanted a status quo, Ayub a change.

By the early 1960s, with weapons flowing from the USA since 1954, the Pakistan armed forces were better equipped and in some areas of armour may even have gained a superiority over India.[40] Ayub might have resorted to force, particularly when a window of opportunity was opened to him in 1962 when India was caught by surprise by the Chinese invasion in the north-east. Ayub refused to take advantage of India's difficulties. In retrospect that was the closest opportunity Pakistan would ever get of recovering Kashmir. The military débâcle and humiliation at the hands of the Chinese left the Indians determined to build up their armed forces. There were moments in the Sino–Indian war when it had seemed that nothing could stop the Chinese from penetrating deep into India. Nehru in desperation had turned to the USA not only for combat aircraft but also for American personnel to keep out the Chinese.[41] Despite lip service to non-alignment, India was now willing to accept aid from any source, irrespective of ideological constraints.

The spectre of India being invaded by a communist power rang alarm bells in Washington and in other NATO Power capitals. On 29 October 1962, Britain began airlifting military supplies to India and shortly afterwards massive supplies also began to arrive from the USA. Before the end of the year Britain and the USA had given military aid to the tune of 120 million dollars. The US engineers helped to build Srinagar–Leh road and made the Leh airstrip operational for combat services. Such was the anxiety in the West about Chinese expansion that in May 1964, the USA announced a further package of military aid of $200 million to India, extending over five years.[42] The USSR, partly motivated by its growing rivalry with China and unwilling to see India pushed too far into the Western camp, also gave more MIG aircraft, and assistance for the manufacture of MIGs in India. Simultaneously, India itself launched a five year defence plan at a cost of Rs. 5000 Crores for modernisation, re-equipment and expansion of the armed forces.[43]

Pakistan was naturally enough perturbed by the arming of India. In Pakistan's view the Sino–India conflict was no more than a border skirmish in which China's limited objectives were accomplished and therefore the armaments supplied to India by the USA and Britain would eventually be used against Pakistan. The Americans, mindful of their Korean experience were convinced that China was an expansionist power and were therefore anxious to help India contain her. While the Americans refused to make their arms supply to India conditional on the settlement of Kashmir, the US Secretary of State, Dean Rusk, together with Duncan Sandys and Lord Mountbatten (now the Chief of the UK General Staff), gently persuaded Nehru to accept a 'low key mediation' or 'good offices' by an American or a Briton in the Kashmir dispute.[44] Between December 1962 and May 1963, Sardar Swaran Singh and Zulfiqar Ali Bhutto, the Foreign Ministers of India and Pakistan respectively, met no fewer than six times. Pakistan was willing to give up its claim to Jammu but was adamant that the fate of the Kashmir Valley could only be decided by a plebiscite under international auspices. But India insisted that the existing cease-fire line, with minor modifications, should be converted into an international boundary and arguing, perhaps with an eye to its anti-communist supporters in Washington and London, that the defence of Ladakh against China depended upon the control of Srinagar. Clearly, some progress had been made. Neither side was now claiming the entire Jammu and Kashmir, and the only sticking point appeared to be the future of the Valley.[45] But what little hope there was of a negotiated settlement was ended by the Kennedy–Macmillan Birch Grove statement of 30 June 1963 which confirmed the US and British military aid to India without any reference to the settlement of the Kashmir dispute.[46] For Pakistan diplomacy this was a turning point: the US fidelity had been tested and the alliance had been found wanting. Once the pressures were taken off from India, she was no longer keen to pursue negotiations which threatened to disrupt the status quo.

The main moral for Pakistan in the aftermath of the Sino–Indian conflict was not to put all their eggs in one basket. Because she had been the 'most allied' ally of the USA, her friendship had been taken for granted. By 1963 Ayub Khan had secured his political position and with his own house in order he proceeded to evolve a new diplomatic stance. Defying the USA he began to move towards non-alignment and closer relations with China and the Soviet Union.[47] China and Pakistan, which had begun making friendly gestures through an oil exploration agreement in 1961 now further strengthened their ties by settling their border question in 1963 and entered into a number of treaties including an air transport agreement whereby Pakistan International Airlines became the first outside airline to operate regular commercial flights to China. The bilateral ties were further cemented when Zhou Enlai visited Pakistan in February 1964 and openly expressed China's support to Pakistan's claims in Kashmir. China soon became Pakistan's staunchest ally.[48]

Pakistan had always been wary of involving the Soviet Union in the affairs of the sub-continent because, unlike the USA, the USSR saw India as the main power in the region and recognised that such a position could not be altered by supplying arms to Pakistan. She had therefore moved closer to India by consigning Pakistan to a position of inferiority. But with the flow of Western arms to India and the resurgence of anti-communist sentiments in India, the old doubts about the bourgeois origins of the Indian leaders began to surface again. Moreover, the USSR was also anxious not to hand over Pakistan to China on a platter. The Russians, rather pragmatically, decided to pursue Pakistan's friendship without alienating India. Besides the USSR's main concern continued to be its rift with China and therefore nothing was to be gained by weakening India. Rather than switching alliance, the Kremlin now donned the role of a mediator.[49]

The changed Russian posture fitted in well with Ayub Khan's diplomacy. Without actually hampering relations with China, he was anxious to pursue a policy of 'flexible equidistance' with both China and the USSR. In August 1963, the Soviets offered a large loan to Pakistan, signed a barter and an air agreement. By May 1964 the Soviets had dropped their earlier unqualified support for India's position on Kashmir at the Security Council. And following Ayub's visit to the Soviet Union in April, the relationship was further cemented: the Soviets offered another loan of $50 million for oil exploration, an agreement to double the Pakistan–Soviet trade in two years and a cultural exchange agreement. For the first time Pakistan–Soviet Friendship became a going concern in Pakistan.[50]

Simultaneously, Pakistan also launched a diplomatic campaign to strengthen its traditional links with fellow Islamic countries. In the summer of 1964, largely at Pakistan's initiative, the plan for Regional

Cooperation for Development (RCD) was launched with Iran and Turkey as the two other partners.[51] Another by-product of the China–Pakistan cordiality was improved relations with Indonesia. This was a notable achievement considering Indonesia was one of India's closer allies in the non-aligned movement. Further diplomatic support also came from the United Arab Republic and Saudi Arabia.

The principal motive behind these feverish diplomatic manœuvrings was to twist India's aims over Kashmir. In the period 1964–5 there was a considerable anxiety in Pakistan that her military advantage which she had built up through alliance with the West would be eroded in the wake of India's massive rearmament both by the West and the Soviet Union. Indeed, by September 1965, when military aid was halted to both countries, the US economic aid to India exceeded six billion dollars compared with three billion dollars to Pakistan. Even though Pakistan had received substantial military aid herself, she was worried because the US economic aid to India had freed India's domestic resources to build up its armaments. It was increasingly argued, particularly by Bhutto, that Pakistan must act before the military balance had tilted in India's favour and the window of opportunity closed forever. It was also argued by the military leaders, particularly General Mohammed Musa, the Commander-in-Chief of the Army, that despite an overall disparity because of India's military build-up, Pakistan still had 'theatre superiority'.[52] In other words, a localised war fought specifically in Kashmir could still be won. All indications were that unless Pakistan acted decisively it would be too late after 1965. Neville Maxwell, the author of *India's China War* and then *The Times* South Asian correspondent, observed the developments from both sides and gained considerable insight, has described the 1965 war as 'Pakistan's India War'. Maxwell gives three reasons for Pakistan's decision to use force: Nehru–Ayub negotiations had dashed Ayub's confidence in securing a negotiated settlement as he began to perceive that as far as India was concerned, Kashmir was a frozen issue and India would not make any tangible concession. Second, diplomatic pressures from Britain and the USA were unlikely to make any difference to India. And finally, Pakistan had to act because Kashmir was vital for Pakistan. As he told Maxwell, Pakistan would have to have Kashmir because of 'blood, water and iron', meaning presumably that the Kashmiris were Muslim, that the headwater of several Pakistani rivers were in Kashmir, and that it was a matter of prestige and pride for the armed forces.[53]

Before the military disparity widened too far and closed Pakistan's last window of opportunity, she must act. In 1965, the moment seemed to be ripe. Pakistan had a quasi-alliance with China, her relations with the Soviet Union were better than ever before (or since), the traditional links with the Islamic countries were cordial and becoming stronger, and despite coolness in relationships with the USA, there was enough residual goodwill to preclude any damaging hostility. The internal developments in Pakistan, India and Kashmir, to which we now turn, also seemed conducive to Pakistan's plans in Kashmir.

The military débâcle in the Sino–India war in 1962, followed by the death of Jawaharlal Nehru who had been India's Prime Minister since Independence in 1947, had left India demoralised, unsteady and confused. Although it showed the great strength of the Indian democratic tradition that despite much foreboding, succession to Nehru was orderly, the new Prime Minsiter Lal Bahadir Shastri, without ignoring his many outstanding qualities, was a compromise candidate and was seen as a caretaker administration.[54] Moreover, the centre-state relationship was going through a strain: the Congress governments in a number of states had been ousted and in the South the Tamils were agitating for some sort of an autonomous Dravidian state. Communal conflicts were rife and adding to the tensions in the sub-continent.[55]

In contrast to India, the political and economic situation in Pakistan in the mid-1960s appeared to be better than they had ever been since independence. Ayub Khan had at least ostensibly legitimised his usurpation of power through a referendum, and the country had been given a constitution and the semblance of democracy. Economically, Pakistan had made rapid progress and her development was hailed as a model for other developing countries.[56] In fact, Pakistan had overtaken India in economic development in the period 1960–65, the fact that such development was at the expense of greater economic inequality was not noticed in the euphoria.[57] The abundance of consumer goods in the shops and massive spending through the development programmes had assured the support of the élites for the regime. The rapid commercial and industrial expansion opened up job opportunities for the educated classes hitherto undreamt of.

'There is a tide in the affairs of men which taken at its height leads to fortune' so Ayub and Bhutto would have probably reasoned. Ayub had no doubts, given the emotional importance of Kashmir to the masses, that if he could solve the Kashmir issue all his past misdeeds would be forgiven and he would be assured of his presidency for life. Bhutto, who combined an unrivalled appreciation of international politics with a streak of ruthless determination, concurred with the military authorities on the necessity of striking before the military balance tilted in favour of India. After the failure of his talks with Swaran Singh, Butto's attitude began to harden. He now rejected the idea of a limited plebiscite in the valley and sought a reversion to a full plebiscite as recommended by the UNO and earlier promised by India. Bhutto consistently worked to arouse popular opinion in Pakistan and brought the Kashmir question into the forefront of the presidential campaign in the winter of 1964–5. To an extent he had upstaged his political master Ayub and thereby cornered him into a position where he would be under pressure to take a strong line on Kashmir. Towards the end of December 1964 Bhutto euphorically promised 'better results in the near future'.[58] Once the expectations were raised to fever pitch, back-pedalling would be a political suicide. Bhutto's motives are not difficult to discern. The ambitious Wadera from Larkhana had calculated his moves shrewdly. If Pakistan could wrest Kashmir from India he would, as the architect of the policy, bask in success and be assured of inheriting Ayub's political mantle. On the other hand if the scheme backfired, the responsibility would be with Ayub. Either way Bhutto's ploy to capture power could not fail. Bhutto's action clearly demonstrated how the Kashmir issue could be used for manipulating domestic politics.

If Pakistan was looking for an opportunity to ignite the Kashmir fuse, the developments within Indian-held Kashmir were indeed propitious. In October 1963 India began moves to do away with Article 370 of the Indian Constitution, which allowed Kashmir special status, and sought to amalgamate Kashmir more fully into the Indian Union, and thereby remove an anomaly which was so much resented by the other states. According to the plans the Sadar-i-Riyasat and the Prime Minister of Kashmir would be downgraded to Governor and Chief Minister of a State, and the integration would be completed by allowing Kashmir to send six members to the Lok Sabha in New Delhi.[59] The gradual erosion of Kashmir's special status under which Kashmir's Constituent Assembly had acceded to India was greatly resented by the Kashmiri Muslims. But with Sheikh Abdullah in prison the lid was kept on firmly. However, trouble flared up from an unexpected quarter.

It was reported on 27 December 1963 that a holy relic – claimed to be the sacred hair of the Prophet Muhammed – was stolen from the

Hazratbal mosque in Kashmir. Whether this was an accident or a deliberate act to escalate communal tension is not certain. Maxwell suspects the hands of the former Chief Minister Ghulan Bakshi.[60] Bakshi, it appears had made a gesture of goodwill towards the so-called Kamraj Plan whereby many prominent Congress leaders were called upon to resign from office to devote themselves to the task of rebuilding the Congress Party which the recent elections had shown to be losing popular support. But to his surprise and regret his resignation was accepted by Nehru. Bakshi, finding the discomforts of being out of office intolerable, had sought to put pressure on the central government by provoking a communal conflict. Whether or not the story of Bakshi's complicity is true, the result was quite dramatic. There was an immediate and widespread rioting which was interpreted in Pakistan as 'the defiant-struggle of Kashmir's four million Moslems to be free'. The story of the uprising was given international publicity by Richard Critchfield, the correspondent of Washington's *Sunday Star*, who accidentally happened to be holidaying in Srinagar. He wrote with some exaggeration:

> After two weeks it is impossible for an outsider even one deeply sympathetic to India, to believe India can continue to hold on to Kashmir ... India's 15-year attempt to win Kashmir is ending in tragic failure.[61]

The sectarian fighting in Kashmir spread and different parts of Pakistan and India were soon engulfed in communal frenzy, thereby demonstrating the popular passion.

The curtain raiser before the actual armageddon was played out in the Rann of Kutch in the spring of 1965.[62] The Rann was a disputed territory and had long been so. The dispute, involving some 3500 square miles of sandy waste, had existed between the province of Sind and the princely state of Kutch. The Radcliffe Award of 1947 gave Kutch to India but did not delineate the boundary and therefore the dispute was transferred to the new states. Pakistan claimed that the boundary ran through the middle of the Kutch, while India maintained that it lay on the northern edge of the Rann. Negotiations had continued intermittently and the local military and police commanders of both countries turned a blind eye to the encroachments and trespassing by the nationals of the other. However, in the aftermath of the deafeat by China, India was touchy about boundary disputes and reacted rather high-handedly. An Indian police officer, only recently posted to the Rann area, set about to seal the border and extend India's control over the disputed areas. Pakistan saw in India's actions a repetition of her action in the north-eastern boundary dispute and realising that militarily she stood on a firmer terrain decided to test out India's strength and determination to fight. Indian troops were seriously disadvantaged. All the local advantages were with the Pakistanis: the Rann was well connected with roads from Pakistan and the border was close to Pakistan's forward positions making it easy to move troops and supplies to the battle-front. Thus when fighting broke out on 9 April 1965, the Pakistanis launched a massive tank attack and had no difficulty in routing the Indian outposts. The Indians, recognising the overwhelming tactical disadvantages, had beaten a hasty retreat rather than losing lives and equipment. Pakistan had demonstrated its operational superiority on a localised conflict. The military victory and the subsequent truce through British mediation which provided for arbitration and a self-implementing agreement to which both sides agreed to adhere. The Rann war also confirmed Pakistan's perception of India's military weakness and the indecisiveness of India's political leadership. What was not appreciated by the Pakistani leaders was that after successive humiliations in 1962 and April 1965, the Indian military had strengthened its position substantially, the retreat from the Rann was tactical and not an evidence of its military weakness or the lack of nerves; and more crucially the Indian public would not tolerate another military humiliation and therefore, the scope for political manœuvre by Lal Bahadur Shastri's government was considerably circumscribed. In other words India would fight with all her might to win a much needed military victory.

Pakistan completely misread India's mood. The euphoric Pakistan leaders had tested India's nerve and having found it wanting, now launched their plans for 'operation Gibraltar' to recover Kashmir. A small committee headed by Bhutto decided on a Rann of Kutch type local campaign confined to Kashmir. Aziz Ahmad, an influential civil servant, convincingly argued that fighting in Kashmir over the disputed cease-fire line (CFL) would remain localised and that India would not dare to antagonise world opinion by attacking on the Indo–Pakistan international boundary. According to the plans Pakistan would send out Mujahiddin and commandos across the porous CFL and, given the enormous popular discontent in Kashmir, their very presence would encourage the Kashmiris to rise in revolt. Pakistan's army would then appear to be seen as coming to the aid of the Kashmiris fighting Indian brutalities. The Kashmir dispute would be back on the table and India would be forced into accepting arbitration with a self-implementing formula akin to the settlement of the Rann dispute.

The whole plan backfired because the two assumptions on which 'Operation Gibraltar' was based proved wrong. Far from the Kashmiri Muslims actually rising in rebellion, they actually apprehended the Mujahiddin when they crossed into Kashmir in August 1965, and handed them over to the Indian authorities. From then onwards the Pakistan high command was divided. Ayub Khan lost his nerve and wanted to call off the plan. However, the hawks in the committee, especially Bhutto, encouraged the Sector Commander, Major General Akhtar Malick to push ahead for the strategically important Akhnour, despite the fact that the expected rebellion of the Kashmiri Muslims never happened. But counsel was divided. Before Akhnour could be reached Ayub removed Akhtar Malick from command and replaced him by his faithful ally, Lieutenant-General Agha Mohammed Yahya Khan. The thrust was at once slowed down. By this time the Indian armed forces fearing that the loss of Akhnour would make Kashmir indefensible, launched 'operation grand slam', which in order to halt Pakistan's armoured thrust into Kashmir and to compel the Pakistani concentration of forces in Kashmir to thin out, actually launched an attack on Lahore and Sailkot. The Indians had refused to confine the fight to Kashmir and had shown little hesitation in violating the international frontier. The second assumption of 'Operation Gibraltar' had also proven wrong. Given Pakistan's lack of depth she could not afford to lose territory in order to accomplish her objectives in Kashmir. The desperate but daring move by India to cross the international frontiers saved Kashmir. Pakistan had to move her forces to protect her own territory. Even though Pakistan fought some heroic battles and her air force virtually knocked off the numerically superior Indian air force, she lost the war because her strategic objective remained unfulfilled.[63] The war which began on 5 September ground to a halt twelve days later as the USA put an embargo on arms and neither side had the resources to fight on their own. Both sides gratefully accepted the UN call for a cease-fire on 17 September.

It was the Soviet resolution in the UNO which resulted in a cease-fire. The Soviet Foreign Minister Kosygin next persuaded the two sides to meet at Tashkent to resolve their differences.[64] In the event, under Soviet prodding, both parties agreed to withdraw their 'armed personnel' to the positions held before the outbreak of hostilities but the Tashkent agreement, like the war itself, did nothing to break the stalemate. It merely restored the status quo antebellum. The two parties had arrived with differing perceptions of what issues would be discussed at the conference.[65] India brought detailed plans for the restoration of the relations between the two countries. It therefore proposed a no-war pact, the resumption of diplomatic relations, freezing the cease-fire line into international boundaries, the release of the prisoners of war, restoration of communication and a cessation of hostile propaganda. But on the crucial question of Kashmir India insisted on a status quo insisting that it was an integral part of India and therefore not negotiable. For Pakistan, on the other hand, Kashmir was at the core of the conflict and all else was peripheral. The failure to obtain a self-executing machinery for the settlement of Kashmir, like the one in the Rann, was immensely disappointing. Both sides returned home disappointed. The Tashkent agreement had no more than ended open hostility which neither country could in any case afford to continue.

The domestic consequences for both Shastri and Ayub were considerable. Shastri had failed to obtain a no-war pact, or convert the cease-fire line into a line of actual control, and yet agreed to give back the territory captured by the Indian troops in Pakistan-held Kashmir. Nor did Pakistan offer any guarantee against infiltration of the Mujahiddins from Azad Kashmir. However, Shastri's sudden death, hours after signing the agreement, spared him the ignominy and criticism. But for Ayub there was no such escape. In fact, Tashkent was the beginning of Ayub's end. Quite predictably Bhutto distanced himself from Ayub, giving credence to the impression that Ayub had signed the agreement against his advice.[66]

The economic consequences of the war were disastrous for both countries. Defence expenditure in both countries soared higher than ever before. India's GNP dropped by 5.2 per cent in 1965–6, rising by a modest 1.1 per cent in 1966–7. The poor economic growth was compounded by the droughts of 1965–6 and the massive increase in expenditure to replenish the loss of weapons and aircraft. For Pakistan the economic crisis was far more damaging. The suspension of the US economic and arms assistance hurt Pakistan's growth rate and enormously increased the burden on foreign exchange resources to purchase arms. But much more ironically, as discussed in the chapter on Pakistan's domestic security, it dried up the development funds and with it withered away Ayub's network of supporters.[67]

INDO–PAKISTAN RELATIONSHIP AFTER THE LOSS OF BANGLADESH

The long-term consequences for Pakistan were even more disastrous. The 1971 Indo–Pakistan war and loss of Bangladesh was a direct consequence of the 1965 war. During the 1965 war East Pakistan was virtually defenceless and lay at the mercy of India. But India, in what proved to be a most adroit diplomatic stroke, left East Pakistan alone and thereby driving home the point that the conflict was with West Pakistan and its military leaders and by implication that the Bengalis had nothing to fear from India.[68] The Pakistan government had been quick to sense India's scheme and organised anti-Indian demonstrations and, as was popularly believed in Dacca, caused its own air force to drop dud bombs (none of which exploded) on the outskirts of the city. Many in East Pakistan began to believe that the fear of India was largely imaginary and thereby undermined one of the principal factors binding the two wings of the country. Robert Jackson perceptively summed up the effect of this war on East Pakistan:

> The war brought home to Bengalis East Pakistan's vulnerability and the impossibility of defending it by military efforts based in the West. Many of them came to see the cause of the war, the Kashmiri question as an exclusively West Pakistan concern: and the complete severance of East Bengal's economic relations with India following the outbreak of hostilities reinforced the sense of economic grievance which was increasingly felt by the East against the West.[69]

Thus, when the war was over the movement for provincial autonomy, which had been on the cards ever since the early 1950s, now gathered momentum under Sheikh Mujibur Rahman's leadership. East Pakistan's movement for autonomy which might develop into a struggle for secession was clearly in the interests of India: it would undermine Pakistan's two-nation theory, weaken her claims to Kashmir and cut Pakistan to size whereby she would not be able to challenge India's predominance. But India played her cards dexterously. An open support for the autonomy movement would have backfired for it would give credence to the Pakistan government's allegations of the Awami League's links with India. In the event, India did not have to do anything. The military junta in Pakistan mishandled the autonomy question in such a way so as to make secession inevitable.

A mass agitation in Pakistan had uprooted Ayub in 1969 and forced his military successor, General Yahya Khan, to agree to hold a general election in 1970 – Pakistan's first since its inception in 1947. The military had conceded an election under compulsion but were as yet far from being reconciled to handing over power to a civilian regime, still less to the Bengalis. Thus when the Awami League, riding on the crescendo of its six-point programme for autonomy, won an absolute parliamentary majority, the junta attempted to negate the popular verdict by launching a military invasion of East Pakistan.[70] India remained aloof and waited for Pakistan to walk into the quicksand. As the Pakistan army unleashed acts of wanton genocide, millions of Bengalis including most of the Awami League leaders and Bengali deserters from the army fled to India. India still did not intervene directly but she provided bases and arms for the Bengalis to fight for their own independence. Between March and December 1971, the Indian leaders worked relentlessly to build up world opinion against Pakistan's atrocities and thereby isolated Pakistan diplomatically. And after further strengthening her own hand by concluding the Indo–Soviet Friendship Treaty in 1971, she bided time for Pakistan to declare war against her.[71] After several weeks of border clashes, Pakistan's air force launched a preemptive attack on India on 3 December. For the third time in their short history the two countries were at war.

The outcome of the war was never in doubt. Pakistan's calculations had gone wrong on all counts.[72] The USA gave some vocal support but, apart from a gesture of moving a task force into the Bay of Bengal, she kept away from getting involved in a South Asian embroglio and, in any case, had her hands full elsewhere; and Pakistan's most reliable ally, China, turned out to be a paper-tiger. Secondly, Pakistan's hopes of continuing the war long enough for her allies to patch up a cease-fire resolution was foiled in the Security Council by the Soviet Union's veto. The Soviets made sure that no cease-fire would take away the victory from India's clutches. Third, the hope of the Pakistani generals that their forces in West Pakistan sectors would be able to make deep inroads into Indian territory which could then be used as a bargain for the evacuation of Indian forces never happened. Pakistan's defence in East Pakistan collapsed without much resitance, and in West Pakistan the Israeli style preemptive strike failed to cripple the Indian air force; and on the ground the forward thrust was halted very early in the war. Mile for mile, the Indians had occupied more territories than Pakistan even in the Western sectors.[73] And finally, Pakistan had grossly underestimated India's military might or the determination of the Bengalis to rid themselves of Pakistan. The US arms embargo of 1965 had hurt Pakistan more than India. Pakistan, while it had received assistance from China, had been far too much dependent on outside countries for weapon supplies. Her military capability, relative to India, had actually declined since 1965.[74] Moreover, in East Pakistan, over a thousand miles away from Pakistan, she was not only surrounded by India on three sides and blockaded on the sea, but she was also operating in a territory where almost every Bengali to a man was not sympathetic to Pakistan.[75] When after a fortnight Pakistan's allies were unable to halt the war, the Pakistan troops already demoralised and left without air cover or hope of reinforcement, surrended without a battle. Unlike the previous wars, this time the result was decisive. Pakistan had failed to wrest Kashmir but had in the process lost East Pakistan and more than half its population. India had proved that military might and successful diplomacy, and not merely legal and moral arguments, were the crucial determinants in international relations.

The atmosphere in Simla where Indira Gandhi and Bhutto met, together with their large entourage was markedly better than six years earlier in Tashkent. The two wars of 1965 and 1971 had shown the futility of attempting to change the status quo in Kashmir. Moreover, in 1972, the power configuration in the sub-continent had been decisively altered. Pakistan had lost her eastern wing and further disintegration in the West had only been averted by India's unilateral cease-fire. Pakistan's external supporters, especially the USA were convinced, for the moment at least, of the futility of tampering with the obvious: no amount of external aid and armaments would enable Pakistan to be a match for India. Even though at the popular level the defeat was seen in Pakistan as the result of the incompetence of the generals, the more sober opinion in the new democratic government was reconciled to the need to bury the hatchet and coexist peacefully.[76]

In India, the euphoria of the victory did not cloud Indira Gandhi's judgement or her future vision for the sub-continent. She perceptively realised that for the first time since 1958 there was a democratic government in Pakistan, as in Bangladesh, and therefore, India would have to be magnanimous if she were to prevent the initiative from slipping back to the generals. On the other hand, the pragmatic Indira realised this was India's long awaited opportunity to settle the disputes once and forever.[77] Indira Gandhi adopted an approach very different from that of her predecessor at Tashkent. Not only did she want restoration of normal relations and a 'no war' agreement but also proposed to grapple the Kashmir nettle through a bilateral arrangement and without the interference of any outside mediators.

Bhutto was on slippery terrain and had to watch his every step.[78] He could not return from Simla empty handed for it would play into the hands of his opponents; but at the same time to concede too many of India's demands would be courting disaster. Not only had Bhutto no cards to play but he also knew all the trumps were in his opponent's hands. There were 93 000 prisoners of war, many of whom were liable to be tried for war crimes; large areas of Pakistan were under Indian occupation and the population at home was restive for a quick settlement so that life could resume again. Not surprisingly, Bhutto did the only thing he could: buy time. He reversed the earlier negotiating position of Pakistan by insisting on a step-by-step settlement. He down-played the prisoners of war issue hoping that the longer they stayed in India the greater would be the international humanitarian outrage and force India to allow them back without any quid pro quo.

While Bhutto played to the gallery and maintained a rigid posture, he was behind the scenes taking a much more constructive attitude. In order to avoid Ayub's fate after Tashkent, Bhutto had taken with him a large and a representative entourage who were actively involved in the negotiations so that no one could accuse Bhutto of a sell-out. He recognised that Kashmir was a lost cause but could not admit so in public by accepting Indira Gandhi's suggestion that the cease-fire line (CFL) be converted into a line of actual control (LAC). Instead, Bhutto advocated leaving Kashmir as it was with a soft border along the CFL to allow the Kashmiris free access in the whole of Kashmir. As a first, Bhutto wanted the 'softening of the cease-fire line', so as to allow trade and easy travel facilities between the people of Kashmir. Bhutto probably envisaged that a united Kashmir which would be autonomous and on friendly terms with both the countries would emerge in a few years.[79] Indira Gandhi shrewdly desisted from pushing Bhutto too far and worked out an agreement which had the appearance of satisfying both sides. Both leaders agreed:

> That the two countries put an end to the conflict and confrontation that have hitherto marred their relations and work for the promo-

tion of a friendly and harmonious relationship and the establishment of a durable peace in the sub-continent so that both countries may henceforth devote their resources and energies to the pressing task of advancing the welfare of their people.[80]

In Kashmir both sides agreed to accept the line of control as existing on 17 December 1971. The new line of control was considerably advantageous to India, giving it the strategically important points of Tithwal and Kargil and thereby making any future attempt by Pakistan to dislodge India from Kashmir extremely difficult.[81] Although Bhutto signed no formal agreement to terminate the Kashmir dispute, he implicitly accepted the partition of Kashmir as a *fait accompli*. In 1974 Bhutto took steps to integrate Azad Kashmir into Pakistan by giving Gilqit Aqency a representation in Pakistan's National Assembly and thereby throwing overboard Pakistan's stand on self-determination for the Kashmiris. Even Bhutto's military successor, Zia-ul Haq has continued to soft-pedal the Kashmir issue. At the same time internal developments in the Indian part of Kashmir were conducive to a peaceful settlement. In 1972, the Government of India withdrew the ban on Sheikh Abdullah's entry to Kashmir when he dropped his demand for a plebiscite. And three years later he not only was back as the elected leader but also renegotiated with Delhi a specially advantageous political and financial position for Kashmir.

The Simla spirit survived the Conference and in the months following the agreement it seemed that the sub-continent had entered a new era. Despite delays, Pakistan and India took steps to normalise relations. Bangladesh agreed to drop war crimes trials, Pakistan recognised Bangladesh, and India allowed the prisoners of war to return to Pakistan. For the first time since 1965, travel and trade was restored between India and Pakistan and Bangladesh. The existence of democratic governments in India and Pakistan, with their socialist commitments, meant that funds would now be channelled away from military wastage to more constructive programmes. Indeed, many observers believed the revolution of rising economic expectations among the electorates of both countries would compel their governments away from barren confrontation to fruitful cooperation. However, the optimism did not last long.

THE NUCLEAR FACTOR IN INDO–PAKISTAN CONFLICT

The deep-seated suspicions which existed between the two states were not altogether wiped out in Simla and the slightest provocation would revive them. In the years between 1974 and 1977 many of the factors which had contributed to the normalisation of relations were either dramatically altered or disappeared altogether. The military was back in harness both in Bangladesh and Pakistan in 1975 and 1977 respectively; even in India, Indira Gandhi declared a state of emergency in 1976 and began ruling in an authoritarian manner. For the military rulers in Bangladesh and Pakistan a posture of hostility and some elements of belligerence in the sub-continent was an obvious way of securing legitimacy. The annexation of the protectorate of Sikkim in 1975, which will be discussed in another chapter, as the twenty-second state of India once again brought to the surface Pakistan's fears of India's imperialistic designs. But what rocked the boat and once again threw the sub-continent's security into a disequilibrium was India's nuclear explosion in May 1974 and the subsequent arming of Pakistan by the USA in the wake of the Soviet invasion of Afghanistan in 1979. To these two issues, we must now turn for the future of South Asian security hinges on them.

Although India has tried to reassure Pakistan that her nuclear explosion was entirely peaceful and that she had no plans to construct a nuclear bomb, Pakistan is understandably cynical.[82] The effect of India's nuclear detonation on Pakistan is in some ways similar to that which happened to the USSR in the 1940s when the USA exploded an atomic device. The feeling in Pakistan, as it was in the Soviet Union, is that unless she too can get her nuclear bomb she could be blackmailed into accepting the dominant position of her rival. Pakistan may be truncated, and might have realistically reconciled itself to the partition of Kashmir, but it is still unwilling to accept the role of an Indian satellite and is therefore pushing forward to acquire a nuclear capability.[83]

India is in a dilemma. If Pakistan succeeds in getting a bomb India would be obliged to follow suit, for until now despite her capability she has refrained from stock-piling a nuclear arsenal. In fact, India even contemplated an Israeli style pre-emptive attack on Pakistan's nuclear installations.[84] But this would only delay the day of reckoning and make Pakistan more determined to get her bomb in first. The other option for India might be to accept Pakistan's acquisition of nuclear weapons and try to counter it by building up her nuclear arsenal. But this would raise threefold problems: first, India's small neighbours, particularly Sri Lanka and Bangladesh would be frightened into drawing in outside allies to counter Indian predominance – a prospect which India dreads and would like to avoid. Second, given that India does not as yet possess long-range delivery capacity her nuclear weapons would be incapable of effective deployment against China because the Chinese military and industrial concentrations are far away from India; and yet India's possession of nuclear weapons would provoke China to target her nuclear missiles on India and would be capable of serious damage to India. Thus India will increase her security risks *vis-à-vis* China for the doubtful prospects of containing Pakistan. And finally, and no less significantly, the US may embargo export of sophisticated technology which would hamper India's plan for modernisation and retooling of its industries. While these are serious constraints, India has never been known to bow down to the dictates of the bigger powers or to be sensitive to its smaller neighbours.

Pakistan, on the other hand, sees the acquisition of nuclear weapons not as a choice but as a necessity for its very survival. Ever since the mid-1960s Pakistan lost hope of acquiring a parity with India in conventional weapons, a fact underlined by the wars of 1965 and 1971. Besides, unlike India, Pakistan depends almost entirely on imports for aircraft, ships, bombs and other sophisticated weapons; and even though well over half of the national budget is siphoned off for defence, she cannot match India's spending power. Therefore, Pakistan, like China, sees nuclear weapons as the only way of defending itself against a much more powerful rival. While few doubt Pakistan's ability or determination to construct a nuclear bomb, she has adopted a deliberately ambiguous strategy. In addition, Pakistan has been helped by three changes which have in varying degrees worked to her advantage: first, the overthrow of the Shah in Iran has left the USA without a reliable ally and therefore Pakistan acquired an added significance; second, with the election of Ronald Reagan as the President of the USA a new wave of cold war has started and the Republican Administration is keen to bolster all possible allies in Asia; and finally the Soviet occupation of Afghanistan, which will be discussed more fully later, has brought the Soviets dangerously near the oil rich Gulf region and thereby raised Pakistan once again as the front line state in the containment of communism. The importance which Reagan's administration attaches to Pakistan is evidenced by the 3.2 billion dollar aid to Pakistan. But as this aid has a 'cut-off' clause if Pakistan detonates a nuclear device, the Pakistani leaders have shrewdly maintained an ambivalent stance on the nuclear issue. Pakistan has reportedly achieved uranium enrichment capability and could arguably make nuclear weapons without actually detonating a nuclear device. The uranium weapons, unlike plutonium, it has been claimed, do not need to be tested.[85] Pakistan's prudent 'bomb in the basement' policy enables it to continue receiving the US aid including the vital F-16 jets and at the same time leaves India guessing and carrying the onus of making the first move.

There are, of course, other factors which constrain Pakistan's race for nuclear weapons. For the last quarter of a century, barring the brief Bhutto interlude, the army has been the dominant force in Pakistan's politics and has enjoyed clear superiority over the two other services, the navy and the air force. The control and deployment of the bomb would logically come under the domain of the air force, but this is clearly not acceptable to the army generals who run the show in Pakistan. While under Bhutto, Pakistan was steaming full-ahead towards the bomb, under the present government the inter-service rivalry will have to be sorted out before Pakistan goes nuclear. Moreoever, the bomb has important domestic political implications. The acquisition of the bomb might make the role of the military redundant and reduce their coercive powers to intervene in politics – a prospect clearly not palatable to the highly politicised army in Pakistan. There is another final constraining factor, Pakistan is closely linked to the Islamic states and once it has acquired a bomb it will find it difficult not to pass on the technology to other Islamic countries, who have financed her research, without the risk of severely damaging relationships. An 'Islamic bomb' would have a destabilising effect on the Middle East, and Pakistani leaders cannot be oblivious of the implications.

CONCLUSION

The rivalry between India and Pakistan is built into the political structure of the two countries and is therefore likely to be durable. The very existence of one seems to threaten the other. India's secular ideology, the presence of large numbers of Muslims in India and the secession of Bangladesh with the help of India are a constant reminder to Pakistan that her *raison d'être* is under threat. Likewise, Pakistan's determination not only to defy India's predominant position in South Asia, but also to assert its claim to be a homeland for Muslims (despite its loss of the Muslim Bengal) is a source of constant provocation. The Kashmir dispute, over which India and Pakistan have fought three major wars, is symbolic of that much deep-seated hostility. Pakistan, as has been suggested above, despite its very strong legal and moral claims to Kashmir, has virtually reconciled itself to accepting to status quo and has even practically abandoned its insistence on the right of the Kashmiris to self-determination. After 1971 Pakistan appears to have *de facto* accepted the Indian proposition to allow the cease-fire line to become the line of actual control which in time would come to be regarded as the international frontier between the two countries. Both countries have taken important steps to incorporate under their jurisdiction the Kashmiri territories held by them. But despite the virtual shelving of the Kashmir dispute, the hostility remains, Pakistan may have been truncated but has not yet been beaten into submission. Pakistan is yet determined to maintain the balance of power in the sub-continent, however lopsided, and has persistently refused to accept India's hegemony.

Pakistan has sought to achieve a balance of power in the sub-continent through two means. First, it has successfully secured the assistance of foreign powers from several different quarters. The fellow Islamic countries in the Gulf and the Middle East have offered both moral and material support and were instrumental in helping Pakistan back on its feet after the traumas of 1971. Pakistan's alliance with the USA was revived, after a somewhat strained relationship in the mid-1970s, when the Soviet Union invaded Afghanistan in 1979. Pakistan, once again became a front-line state in America's struggle to contain communism. For Pakistan it came as a windfall. The US has already provided a military and economic package with over three billion dollars and including the sophisticated F-16 jet fighters. The massive US rearmament of Pakistan, although ostensibly to help Pakistan to resist the communists, has gone a long way to close the growing military imbalance between India and Pakistan. The Soviet intrusion in Afghanistan has alarmed China which has also joined the USA in continuing its support for Pakistan. The cumulative effect has been to replenish and modernise Pakistan's armour to an extent that Pakistan is once again able to challenge India.

Pakistan, however, recognises the vulnerabilities of relying solely on her foreign allies. Past experience has shown that the USA cannot always be depended upon. China has proved herself as a trusted ally but her own limited resources and technology precludes her from matching the Soviet assistance to India. Moreover, Pakistan is aware that despite massive expenditure on defence it can neither catch up on India's superiority in conventional arms nor match India's defence spending capability. Not surprisingly, therefore, Pakistan has moved towards the acquisition of a nuclear bomb. Pakistan's arguments are compelling. She cannot always rely on her allies her hope of achieving conventional military parity with India is virtually ruled out. Under the circumstances the possession of a nuclear device together with a satisfactory delivery mechanism appears to be credible deterrence. Indeed, as argued above, despite the horrors widely shared by the people in India and Pakistan of nuclear proliferation in South Asia, the acquisition of nuclear bombs by both protagonists is only a question of time. Whether the acquisition of nuclear bombs will actually introduce stability in the region through deterrence must remain questionable. But it will certainly not obliterate the rivalry. Hostility and suspicion of each other has been a way of life for generations and is unlikely to disappear easily.

NOTES

1. For an extreme indictment, see Kate L. Mitchell, *India without Fable* (New York, 1942).
2. A. Seal, *The Emergence of Indian Nationalism. Competition and Collaboration in the Later Nineteenth Century* (Cambridge, 1968) ch. 2.
3. Ibid., ch. 7.
4. J. P. Narayan, *Towards Struggle* (Bombay, 1946) pp. 111–12.
5. G. Pandey, *The Ascendancy of the Congress in Uttar Pradesh 1926–34: A Study in Imperfect Mobilization* (New Delhi, 1978).
6. For Nehru's views see S. Gopal, *Jawaharlal Nehru. A Biography* London, 1975), vol. I, pp. 182–3, 224–5, 238–41.
7. A. Roy, *The Islamic Syncretistic Tradition in Bengal* (Princeton, 1983), passim.
8. Ibid., pp. 63–7.
9. See Jinnah's Fourteen Points, Sayeed, *Pakistan: The Formative Phase*, pp. 70–74.
10. A. H. Syed, Pakistan: *Islam, Politics and National Solidarity* (New York, 1982), see esp. ch. 7.
11. L. Ziring, *Pakistan: the Enigma of Political Development* (Folkestone, 1980) pp. 22–5.
12. Ayesha Jalal, *The Sole Spokesman: Jinnah, the Muslim League and the Demand for Pakistan* (Cambridge, 1985) pp. 88–125.

13. See M. Hasan, *Nationalism and Communal Politics in India 1916–1928* (New Delhi, 1979) ch. 8.
14. K. Callard, *Pakistan: A Political Study* (London, 1957) pp. 14–15.
15. See A. Lamb, *The Kashmir Problem* (New York, 1966): S. Gupta, *Kashmir: A Study in India–Pakistan Relations* (London, 1966): J. Korbel, *Dangers in Kashmir* (Princeton, 1954): M. Brecher, *The Struggle for Kashmir* (Toronto, 1953); Lord Birdwood, *Two Nations and Kashmir* (London, 1956).
16. Cited in S. M. Burke, *Pakistan's Foreign Policy, An Historical Analysis* (London, 1973) p. 16.
17. W. H. Morris-Jones, 'The Transfer of Power, 1947', *Modern Asian Studies*, XVI (1982) pp. 1–32.
18. Cited in Burke, *Pakistan's Foreign Policy*, p. 16.
19. Ibid.
20. *Time Only to Look Forward; Speeches of Rear Admiral The Earl Mountbatten of Burma* (London, 1949) p. 42.
21. Cited in V. P. Menon, *The Story of the Integration of the Indian States* (Calcutta, 1961) p. 316.
22. Burke, *Pakistan's Foreign Policy*, p. 25.
23. Cited in Burke, *Pakistan's Foreign Policy*, p. 25.
24. Ibid., p. 26.
25. Ibid., pp. 27–46; Gupta, *Kashmir*, chs 7–9.
26. Z. Khalilzad, *Security in Southern Asia 1: The Security of Southwest Asia* (Aldershot, 1984) p. 116.
27. Burke, *Pakistan's Foreign Policy*, pp. 44–6; Callard, *Pakistan*, p. 308.
28. Burke, *Pakistan's Foreign Policy*, p. 227.
29. Ibid., pp. 201–8.
30. W. A. Wilcox, 'Political Role of Army in Pakistan', in S. P. Varma and V. Narain (eds), *Pakistan Political System in Crisis* (Jaipur, 1972) pp. 31–9.
31. Ibid. p. 36.
32. Gowher Rizvi, 'Riding the Tiger: Institutionalizing the Military Regimes in Pakistan and Bangladesh', in C. Clapham and G. Philip (eds), *The Political Dilemmas of Military Regimes* (London, 1985) p. 203.
33. Gupta, *Kashmir*, pp. 277–82.
34. Ibid., pp. 469–78.
35. T. Ali, *Can Pakistan Survive? The Death of a State* (Harmondsworth, 1983) chs 2–4.
36. Gupta, *Kashmir*, pp. 447–53.
37. M. Ayub Khan, *Friends Not Masters: A Political Autobiography* (London, 1967) pp. 121–9.
38. Ibid., pp. 127–8.
39. Gupta, *Kashmir*, pp. 343–7, see also C. S. Jha, *From Bandung to Tashkent: Glimpses of India's Foreign Policy* (London, 1983) pp. 174–97.
40. O. Marwah, 'India's Military Intervention in East Pakistan', *Modern Asian Studies*, vol. XIII (1979) pp. 549–80, esp. pp. 552–3.
41. S. Gopal, *Jawaharlal Nehru: A Biography* (London, 1984) vol. III, pp. 228–9, 251–5.
42. Burke, *Pakistan's Foreign Policy*, p. 289.

43. O. N. Mehrotra, 'Arms Build-up in Pakistan and India', in S. Chopra (ed.), *Studies in India's Foreign Policy*, pp. 213–39: for a historical background to the growth of the army, see S. P. Cohen, *The Indian Army: Its Contribution to the Development of a Nation* (Berkeley, 1971).
44. Burke, *Pakistan's Foreign Policy*, p. 283.
45. Ibid., pp. 278–84.
46. Ibid., p. 289.
47. G. Singh, 'Pakistan's China Policy: Causal Considerations 1960's', in S. Chopra (ed.), *Perspectives on Pakistan's Foreign Policy* (Amritsar, 1983) pp. 272–87.
48. Burke, *Pakistan's Foreign Policy*, p. 294.
49. D. Kaushik, 'Soviet Union's Pakistan Policy: A Survey and Appraisal', in Chopra (ed.), *Perspectives on Pakistan's Foreign Policy*, pp. 240–71; S. Mansingh, *India's Search for Power*, p. 199.
50. Burke, *Pakistan's Foreign Policy*, pp. 299–302.
51. Ibid., pp. 305, 307.
52. Most of the information below on the Rann of Kutch War and the 1965 Kashmir War is based on Neville Maxwell's talk on 'Indo–Pakistan Rivalry' at the Seminar on 'India's Security: Past, Present and Future' at St Antony's College, Oxford, 27–29 October 1984. I am deeply indebted to him.
53. Maxwell, 'Indo–Pakistan Rivalry'.
54. R. Kothar, *Politics in India* (New Delhi, 1970) pp. 309–11.
55. S. Gopal, *Jawaharlal Nehru: A Biography* (London, 1979) ch. 12.
56. F. Gustav Papanek, *Pakistan's Development: Social Goals and Private Incentive* (Cambridge, Mass., 1967).
57. K. B. Griffin and A. R. Khan (eds) *Growth and Inequality in Pakistan* (London, 1972) pp. 1–21.
58. L. Ziring, *The Ayub Khan Era: Politics in Pakistan 1958–69* (Syracuse, 1971) p. 53.
59. Burke, *Pakistan's Foreign Policy*, p. 319.
60. Maxwell, 'Indo-Pakistan Rivalry'.
61. *Sunday Star*, 19 January 1964.
62. Maxwell, 'Indo–Pakistan Rivalry'; see also L. Ziring, 'The Rann of Kutch Arbitration', in Masuma Hasan (ed.), *Pakistan in a Changing World. Essays in Honour of K. Sarwar Hasan* (Karachi, 1978) pp. 140–57.
63. A. Lamb, *The Kashmir Problem* (New York, 1966) p. 123.
64. S. Chopra, 'Kashmir as a Factor in Indo-Soviet Relations' in S. Chopra (ed.), *Studies in India's Foreign Policy*, pp. 101–31, see esp. 125–9.
65. C. S. Jha, *From Bandung to Tashkent*, pp. 223–50.
66. Ziring, *The Ayub Khan Era*, p. 84.
67. Ibid., pp. 86–113.
68. Sheikh Mujibur Rahman in a conversation with the author.
69. R. Jackson, *South Asian Crisis: India–Pakistan–Bangladesh* (New Delhi, 1978) p. 21.
70. L. Ziring, 'Militarism in Pakistan: The Yahya Khan Interregnum', in W. H. Wriggins (ed.), *Pakistan in Transition* (Islamabad, 1976) pp. 198–232.
71. S. Mansingh, *India's Search for Power* (New Delhi, 1984) pp. 213–25.

72. Ibid., p. 223.
73. Ali, *Can Pakistan Survive?* p. 95.
74. L. Ziring, 'Dissonance and Harmony in Indo–Pakistans Relations', in S. Chopra (ed.) *Studies in India's Foreign Policy*, pp. 248–9.
75. Ali, *Can Pakistan Survive?* p. 95.
76. Chopra, 'Indo-Pakistan Relations', p. 178.
77. Mansingh, *India's Search for Power*, p. 226; M. Ayoob, *India Pakistan and Bangladesh* (Delhi, 1976) pp. 75–87.
78. Chopra, 'Indo–Pakistan Relations', pp. 174–5.
79. *New York Times*, 3 July 1972, cited in Chopra, 'Indo–Pakistan Relations', p. 178.
80. S. Kumar, *Documents on India's Foreign Policy, 1972* (New Delhi, 1975) pp. 147–9.
81. Mansingh, *India's Search for Power*, p. 229.
82. K. K. Pathak, 'Nuclear Policy of India: Restated', in Chopra, *Studies in India's Foreign Policy* pp. 382–94.
83. B. M. Kanshik, 'Pakistan's Nuclear Policy', pp. 374–92, and T. C. Bose, 'Nuclear Proliferation: Pakistan, A Case Study', pp. 393–407, in Chopra (ed.), *Studies in India's Foreign Policy*.
84. *India Today*, 15 February 1985, p. 18.
85. R. R. Subramaniam, 'The Nuclear Factor in South Asian Security', in paper presented at a seminar on 'India's Security' held at St Antony's College, Oxford, 27–29 October 1974.

5 The Role of the Smaller States in the South Asian Complex

GOWHER RIZVI

Having examined how the Indo–Pakistan rivalry provides the major dynamics for the South Asian security complex, we now turn to investigate the role of the smaller states of South Asia in shaping the security complex. Here we are primarily concerned with fear interrelated questions: the first is to ascertain the role of the smaller states in the rivalry between India and Pakistan and to determine if they make any tangible difference in that conflict; second, we examine the importance of some of the smaller states as buffers between South Asia and its neighbouring complexes and to see if they continue to perform their historical role for which they had been set up; third, we assess the role of these states in facilitating penetration by external powers of the complex; and finally, of course, to determine whether the character of the complex preserves or erodes or enhances the independent status of the smaller states.

INDO–PAKISTAN AND THE SMALL STATES

Before proceeding further it is necessary to clarify the concept of small states. In our discussion of small states we have included Bangladesh, with an estimated population of over 100 million, which is more than that of Pakistan, at one end of the spectrum, and Bhutan and Sikkm (until 1975) with populations of under one million at the other end. In any other context Bangladesh would be treated as a large country but situated as it is next to India, it cannot but feel overawed and in fact psychologically considers itself a small state.

Even the claim to treat Nepal and Sri Lanka as small states would seem a bit absurd had these two states been situated in Africa or South America. Burma, although it was part of India until 1935, is culturally and politically more a part of the South-east Asia complex and therefore we skirt round it; the place of Afghanistan in a South Asian complex remains ambiguous; however, we have already noted its rivalry with Pakistan and will focus in this chapter on the extent to which it impinges on South Asian security, particularly if its buffer role were to be completely eroded by the Soviet occupation of the country.

With one in every five persons in the world living in South Asia, and a land area covering nearly four and a half million square kilometres (4 468 000 sq. km), the region is vast and complex. Although South Asia has clearly defined external frontiers – the Himalayas in the north and the peninsula surrounded by the Indian Ocean – the internal political geography follows no clear lines of demarcation. This is not surprising as the political boundaries are neither the product of history nor geography, but determined artificially on lines of religion and ethnicity. In an ancient civilisation like India where the populations are inextricably interwoven, the boundary demarcations invariably cut across communities and tribes. Naturally enough the boundaries between India, Pakistan and Bangladesh are drawn along religious lines and are the most contentious ones. Moreover, the three major river systems – the Indus, the Ganges and the Brahmaputra – by cutting across the boundaries of these three states further exacerbate the tensions between them.[1]

Although South Asia, with its well defined external boundaries, constitutes a coherent region there is as yet very little linkages among the states – a legacy of colonial rule. When the British withdrew from the sub-continent they transferred power to four successor states – India, Pakistan, Burma and Ceylon – each of whom developed close links with Europe rather than among themselves. Ceylon had air and sea links only with India; Nepal, Pakistan and Sikkim until recently only with India; and Afghanistan only with Pakistan. Such poor communications, together with lack of cultural contacts or even exchange of newspapers, meant that each of these states remained fairly isolated within the region. Indeed, even the excellent road and railway links that existed between India and Pakistan were allowed to fall into disuse in the aftermath of the partition.

Despite these complexities South Asian is probably the best defined subsystem in the contemporary international system and offers a unique opportunity for analysing the role of the smaller states in the regional security complex. Even though the size, population and resources of India makes her pre-eminent in South Asia, she is certainly not predominant in the hierarchy of the states. The lack of integration and only a minimal amount of interdependence in inter-state relations rules out the classification of the system as hierarchical.[2] The best way to define the South Asian structure is therefore to view it as bipolar – even though perhaps it is a lopsided bipolarity – with India and Pakistan as the two poles and the smaller states clustered around them. The lopsidedness of the bipolar structure is further emphasised by the fact that Pakistan, despite often cordial relations with the smaller states, stands virtually aloof and is isolated.[3] India, with some minor reservations, looms as the centre-piece of the South Asian subsystem. The smaller states except for Afghanistan, either voluntarily or because of geographical compulsions, came within the orbits of India's influence. While few of the six South Asian states have common borders with each other, they all (again except for Afghanistan) border with, or are adjacent to, India. Nevertheless, Pakistan despite being truncated in 1971 and with a population which is one-tenth of India's, has persistently refused to acknowledge India's supremacy.[4] It is able to back up its claims militarily and thereby give the structure a bipolar character. Pakistan's ability to defy India's dominance also gives the smaller states the scope to chart a more independent course by playing the two regional powers one against the other.[5]

But being at the centre brings both advantages and problems for India. Five of the six states have borders with India, and this has resulted in inevitable complications since all these states are in their infancy and the boundaries are as yet not firmly settled. India also shares ethnic, religious and cultural affinities with all its neighbours which is both a source of strength and discord. The Tamil insurgents in Sri Lanka put considerable strain on Indo–Sri Lanka relations as the Tamils in Karnataka would like to offer assistance to their brethren across the Palk Straits;[6] the Bengalis in West Bengal and Bangladesh share much in common but under the shaky military regime in Bangladesh, deliberate attempts are made to provoke suspicion between the Bengalis across the border.[7]

The close proximity between India and the smaller states has resulted in in-built tension and conflicts. First, there are the cultural and ethnic linkages which defy the artificial border demarcation and thereby arouses fears of Indian cultural imperialism in smaller states.[8] Secondly, several of the smaller states are poor even by sub-continental standards and therefore there is a constant flow of illegal migrants into India where job opportunities are marginally better and this arouses considerable resentment among the local inhabitants, as for example, of the Assamese against the Bangladeshis; or the Northern Biharis and West Bengalis against the Gurkhas.[9] Thirdly, there is the fear, real or imaginary, of irredentism. The Sinhalese accuse the Indian Tamils of encouraging irredentism among the Tamils living in Sri Lanka.[10] Fourthly, India accuses Bangladesh of according support to Mizo guerrillas and similarly Pakistan for encouraging Sikh extremism.[11] And finally, of course, there are dissonances because India's secularism, democracy, federal principles, and linguistic autonomy are at variance with her smaller neighbours.[12] India and the smaller states impinge on each other's politics, and one cannot ignore the other.

The ability of the smaller states to adopt an independent stance in their international dealings is to a large extent both encouraged and circumscribed by India's international outlook and strategic considerations. Although we have noted that India's claim to the leadership of the South Asian hierarchy is disputed by Pakistan, India nevertheless regards the sub-continent as her area of influence. In a limited sense India has inherited security perceptions from British India. Such perceptions involved the preservation of the integrity of the Indian Empire by creating a series of buffers – Tibet, Sikkim, Nepal and Afghanistan – so as to isolate India from the Chinese and Russian empires, the second objective was to keep away any foreign interference in the region by making the Indian Ocean into a British lake.[13] Even though India lacks Britain's imperial might of the nineteenth century, her security perceptions are not far different: the buffers must be maintained and outside powers kept out from the sub-continent. In the changed situation of the twentieth century, both technological and in the international system itself, India's ability to accomplish its objectives is beset with difficulties. The first difficulty arises from hostility with Pakistan and the latter's ability to draw in outside powers to her support, and thus compel India to call in USSR into the region to balance Pakistan's foreign allies;[14] and the second arises from the fact that the sleeping dragon in the north has not only shaken off the opium addiction but has become a potential superpower and actually occupies the buffer state of Tibet, thereby laying open India's northern frontier.[15] These two factors not only colour India's security analysis but also impinge heavily on her relationships with the smaller states.

India has been extremely reluctant to allow the growth of any regional organisations or evolve a pattern of multilateral relations in the region. This is partly because India has a complicated set of relationships with each of the smaller states, in some cases based on what is euphemistically called a 'special relationship'.[16] India's preferred mode of diplomacy is bilateralism because at the one-to-one level, India's size and position ensures that her interests are not ignored by her smaller neighbours. Whereas in a multilateral relationship India fears that the smaller states may gang-up on her and extract more concessions than she is willing to make. But on the other hand, some of the smaller states also have a preference for bilateralism. This enhances their status as independent states and enables them to treat with India on a basis of equality, at least in theory.[17] But with the advent of the Rajiv Gandhi regime, there appears to be a significant shift in policy and New Delhi is showing greater interest in multilateral forms of regional cooperation.

BANGLADESH

Bangladesh, which came into being in December 1971 by successfully seceding from Pakistan, has added an important element in the South Asian diplomatic configuration. With a population of over a hundred million, Bangladesh can hardly be described as a 'small state', but being situated next to a giant neighbour, the Bangladeshis, psychologically at least, demonstrate all the characteristics of a small state. Indeed, Bangladesh's economic poverty, an extreme dependence on foreign aid for mere survival, and a weak political system dominated by the army, has added to the country's vulnerability and thus precluded her from playing a decisive or even a balancing role in the region.[18] Yet Bangladesh is neither a Bhutan nor a Sikkim and cannot be easily brushed aside. By using her Islamic links, and by developing closer relations with Pakistan, Bangladesh has successfully counter-balanced the threat of Indian hegemony – real or imagined.

Despite a gruelling poverty Bangladesh is a unique country. It is one of the few Afro–Asian states which can claim to be a nation-state. The population is almost entirely homogeneous: barring a few small tribal groups, the people are all Bengalis who share a common racial and cultural origin; and nearly 90 per cent of the population is Muslim. The experience of the liberation war united the Bengalis and reinforced this distinctive identity.[19] The Bengali nationalism was avowedly secular but there was a discernible tension. Even though the Bangladeshis had rejected Pakistani nationalism based on ties of religion and secularism became one of the pillars of the new state's ideology, many Bangladeshis, perhaps somewhat paradoxically, saw themselves as both Bengalis and Muslims.[20] To an extent this assertion of a dual identity was the reflection of a historical anxiety and perhaps an unstated feeling of social inferiority *vis-à-vis* the cultural superiority of the Bengali Hindus of West Bengal. In the early years of Bangladesh such a dichotomy did not particularly matter because religion was divorced from politics and the Awami League government of Sheikh Mujibur Rahman vigorously pursued a secular ideology. However, with the assassination of Sheikh Mujib and the coming of military rule the whole question of Bengali nationalism was once again thrown into the melting pot.

The military regime lacking in popular legitimacy, and unable to address itself to the economic and political problems confronting the state, resorted to the one convenient card they possessed: Islamic revivalism. Once again there was a resurgence of fundamentalist sentiments and an anti-Indian venom.[21] It not only distracted popular opinion from real issues but was also a useful instrument for discrediting the Awami League and at the same time rehabilitating in politics those pro-Pakistani Bengali leaders who had been electorally discredited for collaborating with the Pakistani army and opposing Bangladesh's quest for freedom, and who were now back in the political arena. Their trumpeting the call of Islam became a basic instrument in their strategy for survival. The 'Bengalis', through a constitutional amendment, were redesignated as 'Bangladeshis' so as to give Bangladesh nationalism a territorial identity and, more importantly, to distinguish them from the Bengalis in West Bengal.[22] It is against this background of domestic crises in Bangladesh that one must examine Indo–Bangladesh relations and Bangladesh's role in South Asia.

The emergence of Bangladesh had aroused much expectation and great hopes for the future of South Asia. Not only was it believed that Pakistan had ceased to be a credible military threat to India, but also that the secession of Bangladesh had demonstrated the fallacy of Pakistan's nationalism based on Islamic ties. More significantly, Bangladesh's independence had been made possible by India's decisive intervention and energetic diplomatic manœuvres to secure the new state's recognition. India had demonstrated its goodwill and sincerity of purpose by withdrawing its forces from Bangladesh ahead of schedule and in the first few years had pumped enormous economic aid, despite its own pressing needs, to restore Bangladesh's war-torn economy.[23] India and Bangladesh concluded a treaty of friendship, along the lines of the Indo–Soviet treaty, which promised to be mutually beneficial to both sides. While Bangladesh would be free to maintain what armed forces it chose, it was freed from any fears of invasion, and therefore of the need for wasteful extravagance on large standing armed forces. The relationship was further cemented by a common outlook and a shared concern of the two states. Both countries were committed to parliamentary democracy, secularism and socialism, in foreign policy both supported non-alignment and an abhorrence of any foreign power's involvement in the sub-continent. While Bangladesh reciprocated the Soviet Union's assistance at the United Nations during the liberation war by forging especially close relations, it cast its net wide and accepted aid from both the East and the West. India, however, remained Bangladesh's closest ally and helped to nurse her through the early years.[24]

The happy and mutually beneficial relationship was dramatically disrupted by the assassination of Sheikh Mujib in August 1975. The Indian leaders were, of course, horrified by the brutal murder of Sheikh Mujib and his family and could not look with equanimity upon the military rulers, but refrained from being drawn into the internal affairs of Bangladesh. There were a number of disputes between the two countries but none of such a magnitude or urgency as to upset the existing friendly relationship. But, as suggested above, the necessity to divert public attention prompted the military rulers to articulate anti-Indian sentiments. In the changed political ambience even minor irritants developed into major crises and, not surprisingly, India refused to be as pliant or accommodating as it had been during Sheikh Mujib's time. The burgeoning middle-class of Bangladesh had always resented competition from the Indian traders and now began to clamour for 'protection'.[25] The press played up the supposed havoc that was being caused by the smugglers who drained Bangladesh's economy but it was conveniently ignored that businessmen, smugglers and border forces on both sides were milking the profit. And with the re-emergence of the more fundamentalist elements in politics and government, the cultural intercourse between the two Bengals came under severe attack. Once again, as in the days of Pakistan's military rule, there were embargos on Indian films, music, books and newspapers.[26]

As tensions mounted in Indo–Bangladesh relations some of the problems confronting the two states which were earlier being handled in an amicable environment now assumed a new gravity. The situation was further aggravated by a large number of pro-Mujib forces who had fled from Bangladesh and sought refuge in India from where they hoped to launch a movement against the military rulers. Although India did not give any effective help to the pro-Mujib groups, at the same time it could not agree to hand them back to the Bangladesh authorities.[27] Nor was Bangladesh's demand particularly reasonable. In 1971 India had sheltered nearly ten million Bengalis against Pakistani atrocities; it could scarcely refuse the same to political dissidents who, if repatriated to Bangladesh, might even be killed.

One such dispute concerned the Farakka Barrage which had important economic implications for both countries.[28] The dispute originated in the early 1950s and continued after Bangladesh independence. The problem arises because the rivers in the old delta, covering parts of West Bengal, are moribund and the Hoogly river therefore clogs up with silt. As a result the Calcutta port is in serious danger of being rendered inoperative. As early as the 1850s the engineers had suggested a barrage at Farakka to increase the flow of water through the Hoogly river and thus flush out the silt from Calcutta port. The additional water would also ease the problem of Calcutta's water supply and improve drainage. The plan was revived in the early 1960s and water began to be diverted from the Ganges into Bhagirathi in 1975. Mujib's government was aware of the dangers to Bangladesh's agriculture by the decrease in water supply during the dry season but shrewdly recognised that it was a 'technical' problem which could only be resolved through mutual goodwill and cooperation. Not surprisingly, pending a final solution of the dispute, Mujib was able to persuade India to allow sufficient water for Bangladesh's requirements.[29] However, the military regime, not averse to finding an anti-Indian grievance began to assert, as Pakistan had done earlier, that the Ganges was an international river and therefore Bangladesh, as a lower riparian state, had a right to her full share of water. But India maintains that since 2036 kilometres of the 2177 kilometres of the Ganges flows through India, she as the upper riparian state is entitled to as much water as it requires.[30] In addition, to the legality of the question, there are also important economic considerations. Bangladesh claims that it would affect 11.5 million or about 32 per cent of the total cultivable land of the country and the estimated loss of food grains up to December 1979 was 3683 million takas.[31] India, on the other hand, stands to lose her her most important eastern port. The dispute still continues. Bangladesh had misjudged the intensity of India's determination to hold on to its claim. After an abortive attempt to involve the United Nations, Bangladesh has resumed bilateral negotiations. The Farakka has become the Kashmir of Bangladesh's grievance against India.

There are several other minor irritants in Indo–Bangladesh relations but in the existing atmosphere they have been exaggerated out of all proportion. There are frequent boundary disputes largely due to constant change in the course of the rivers or the appearance of tiny islands called *chars*, which are caused by fluvial deposits on the river beds. Under Mujib the undefined boundaries were delineated with ease in 1974 as both sides gave up small territories which it was felt could be better and more conveniently administered by the other. However, the Indians were less obliging in 1979 over the Muhirir Char.[32] Moreover, in 1981 the Indians occupied Purbasha Char (the New Moor Island) near Sundarban despite strong claims to it by Bangladesh. In the latter the patch of mud was mixed with prospects of oil.[33] Both Bangladesh and India have also accused each other of supporting dissidents. India is blamed for abetting the Chakma Shanti Bahini who are fighting the Bangladesh government for greater autonomy for the tribal people of the Chittagong Hill Tracts;[34] and Bangladesh is suspected of giving sanctuary to the Mizo National Front.[35] The lack of goodwill and tensions between the two states is best illustrated by India's erection of barbed wire along the Indo–Bangladesh border.

Although Bangladesh has been a source of considerable pin-pricks to India, it is scarcely in any position to influence, let alone alter, the South Asian security complex. Her rapidly expanding army is fully occupied in domestic affairs and can in no way pose a threat to India either on its own or in alliance. Nevertheless, Bangladesh has not spared any opportunity to enhance its independent status by counterbalancing India's dominant position. Despite bitter memories of massive genocide and brutalities by Pakistan's army, Bangladesh has taken steps to normalise relations with Pakistan largely in the hope of using it as a counterpoise to India.[36] Without actually discarding secularism as a state ideology, Bangladesh has been asserting its Islamic identity and forging closer links with other Muslim countries. Indeed, the Constitution has been amended to include a novel directive requiring the state to develop closer bonds with the Islamic world. As a part of its strategy for distancing from India, Bangladesh's relations with the Soviet Union have also become lukewarm.

By contrast there is increasing contact and cordiality with the USA and China.[37] Another prop in Bangladesh's quest to chart greater independence is its active role in international organisations and its enthusiastic support for South Asian Regional Cooperation of which it was the original proponent.[38] By insisting on multilateral ties, Bangladesh hopes to use the collective strength of the smaller states to counter Indian hegemonic aspirations. But Bangladesh's ability to counteract India's predominance is limited by what has been described as 'God given and nature-given fact'. Surrounded by India on three sides, separated from Pakistan by nearly a thousand miles of Indian territory, having no common boundary with other South Asian states (except for Burma), Bangladesh's ability to organise an anti-Indian alliance is extremely circumscribed.

SRI LANKA

Sri Lanka, situated next to India, which is nearly fifty times larger and more numerous, maintains uneasy, and despite ups and downs, friendly relations with India.[39] The size and geographical proximity of India has left an important imprint on Sri Lanka's outlook and her strategy for survival as an independent state. At the nearest point Sri Lanka is separated by a mere twenty miles of the Palk Straits; and it is a long way from any other country. Geographic compulsions and historical legacies largely explain the island's precarious quest for security. It feels intimidated by its giant neigbour both physically and culturally. Her vulnerability is exacerbated by a persistent conflict between the two main ethnic groups. Although most of the inhabitants of Sri Lanka have originally migrated from India, there is nevertheless a deep-seated animosity between the Sinhalese speaking Buddhists who constitute the vast majority of the population, and the Tamil minority, who are mainly Hindus. But because of close cultural and sentimental ties that exist between the Tamilus living on both sides of the Palk Strait, the consequences of the Tamil–Sinhala conflict cannot be confined to the limits of the islands. They inevitably impinge on Indo–Sri Lanka relationships.[40]

The Tamils who number nearly three million are mainly concentrated in the northern and eastern parts of the island. They originally migrated from southern India during the nineteenth century to work on tea plantations, and have lived there for generations. After Sri Lanka became independent in 1948 it passed a series of laws which discriminated against the Tamils. The two Citizenship Acts (1948 and 1949) were excessively discriminatory: by requiring a proof of either the father or the grandfather being born in Sri Lanka they virtually ensured that very few of the Tamilus could actually establish their claims. It will be recalled that the ordinance for the registration of births did not come into force until 1895, arrangements for registrations were begun in 1897, the birth register did not mention the names of the children until 1908, and in fact the birth registers were not even available in many parts of the country until 1920.[41] Consequently, and not surprisingly, only 134 188 out of 824 430 who applied to be registered as citizens were successful. Two years later when the franchise list was drawn up, the vast majority of the Tamils found themselves automatically disenfranchised. The Sri Lankan government claimed that Tamils were Indians and therefore ought to be repatriated even though many had been living there for generations. The Indian government, on the other hand, disclaimed any responsibility. The result was that about three-quarters of a million Tamils were rendered stateless.[42]

The Tamil problem has gradually worsened over the years. The Sinhalese constitute over 70 per cent of the population, but suffer from a curious minority complex.[43] They are afraid of being culturally swamped by the Tamils, who although a minority in the island are numerically vastly superior if the Tamils living on the Indian mainland are also taken into account. This fear has caused the Sinhalese nationalism – Sinhala only – to be aggressively exclusive. Numerous laws have been passed by the Sinhalese controlled governments to preserve the Sinhala culture with almost a total disregard to the Tamil sensitivities. Fearful of an imaginary ascendancy, the government has enacted numerous laws which far from safeguarding the interests of the minorities, actually even denied equal opportunities to the Tamils in education, employment, trade and political participation.[44] While the cry for cultural safeguards provide the ostensible motives for the blatant intolerance of the Sinhalese, the real reasons may be explained by the fact that the Tamils are industrious, hardworking and, despite both their numerical inferiority and discriminatory policies aimed against them, are doing much better educationally and economically than their numbers would warrant.[45] At first the Tamils, who are mainly concentrated in the north and east, sought provincial autonomy to safeguard their own particularist needs. But as violence broke out, with mounting savagery on both sides, the Tamils – both the Sri Lankans and Indian born – formed the Tamil United

Liberation Front (TULF) to demand a separate homeland for themselves. The periodic communal frenzy and massacres of the Tamils by the government security forces pushed the Tamils into open insurgency and the Tamil liberation forces are now fighting a guerrilla war to achieve their goal for Eelam – a separate Tamil homeland.[46] Both India and Sri Lanka are anxious to prevent further escalation in the savage reprisals between the Tamils and the Sinhalese, but their ability to control events is circumscribed by their domestic politics and international constraints.

President Junius Jayewardene's government is in a dilemma. The Tamil insurgency has now reached a point which defies any easy solution. The Sinhalese intransigence has pushed out the Tamil moderates, leaving the initiative in the hands of the extremists. The 'Tamil Tigers', as the liberation force is called, will not easily give up its demands for Eelam when it is within its grasp. It is arguably still possible that the insurgency can yet be halted by conceding the Tamil demands for organising the Tamil majority districts in the North-East into an autonomous province. It would still be short of Eelam but by implicitly recognising the right of a Tamil homeland and letting them manage their own affairs, it might persuade the majority of the Tamils to accept this compromise. Most Tamils are not secessionists but have been pushed into extremism by the atrocities of the government forces and therefore it is still conceivable that with goodwill and magnanimity the drive towards secession may be averted. But such a bold concession is as yet unacceptable to the Sinhalese majority. It appears that no government which offers such a concession would have much hope of surviving in Colombo.[47]

The alternative to a political settlement is to combat the Tamils through force. Even if Colombo had the will, it would appear to be logistically impossible. The Sinhalese security forces are ill-equipped, undisciplined, bitter and demoralised. Despite technical assistance and training by British and Israeli defence advisers, the recent performance and conduct of the government forces has left much to be desired. They have indulged in wanton savagery and genocide of the Tamil civilians but have made no impact on the guerrillas.[48] The liberation forces, although much smaller, are by contrast well armed and trained, and ideally suited for combat in the guerrilla terrain of north-eastern Sri Lanka. Besides having support from the local populace, the liberation forces operate from safe bases across the Strait in Tamil Nadu.[49] Moreover, it does not appear that Sri Lanka can enlist any effective outside support to offset India which would be necessary were she to opt for a complete military solution to the problem. Distance precludes Pakistan; besides Pakistan would stand to gain little by alienating India. There have been vague suggestions that the USA might be lured in by offers of base facilities at Trincomalee. But as yet the USA, already comfortably entrenched in Diego Garcia, has shown little enthusiasm except for supplying some weapons. Besides a direct involvement could be counter productive. India, helped by the USSR might then increase its support for the insurgents. Being so near, India is in a better position to alter the arms balance. Moreover, Washington is unlikely to commit itself significantly without a firm indication from the Sri Lankan government that it would agree to a political solution. The memory of Vietnam is still too fresh for Washington to step into another Asian civil war.

A possible solution might lie in New Delhi.[50] India might be able to persuade the Tamils to halt fighting by withdrawing the 'safe bases' in Tamil Nadu and cutting off the arms supply. But a precondition for this would be a firm commitment by Sri Lanka to a political settlement with the Tamils. However, at the moment this option is precluded by repeated massacre of the Tamils and a widespread sympathy for the Tamils, especially in southern India. The situation is further complicated by the differing attitudes of the central government in New Delhi and the state government in Tamil Nadu. The Dravida Munnetra Kazhagan (DMK) government in Tamil Nadu has been sympathetic towards the Tamil liberation forces and is unlikely to withdraw its backing merely at the behest of New Delhi.

The Government of India is caught in a dilemma. India cannot rule out Sri Lanka's ability to secure US penetration in the region. It is understandably afraid that a more direct help to the Tamils might push Sri Lanka to lean on the USA and thereby endanger India's interest in the region: India cannot tolerate the presence of any foreign power in Trincomalee which guards the Bay of Bengal. Perhaps a more important factor in restraining India, notwithstanding her role in Bangladesh in 1971, is her reluctance to be seen to be interfering in the internal affairs of a small neighbour. Such interference might frighten Bangladesh, Nepal or Bhutan into seeking greater assistance of outside powers, such as the USA, the USSR or China – the prospect of outside powers interfering in what India considers her sphere of influence is immensely detestable to India.

At the same time India cannot go on walking the tight-rope. The plight of the Tamilis, particularly the reports of ruthless genocide by the government forces is becoming an emotive issue in India's domestic politics, especially in the south, which the government cannot go on ignoring. Despite Rajiv Gandhi's reluctance he might be forced into a quick 'rescue and retire' operation.[51] According to this scenario Indian forces would be deployed to throw out the Sri Lankan security forces from the Tamil majority regions and then immediately withdraw after calling in a UN or some other international peace-keeping force. But understandably, with an eye to the effect this would have on her other neighbours, India is unlikely to intervene except as a last resort. It is possible that as the civil war bites into the Sri Lankan economy the government in Colombo might become more amenable to a political solution. At the moment there are scarcely any grounds for optimism.

Until about the sixteenth century all the invaders of Sri Lanka came from the Indian mainland.[52] Not surprisingly Sri Lanka has charted its steps by carefully trying to balance her independent status with the susceptibilities of her giant neighbour. On important issues of international relations Sri Lanka has sought to harmonise her actions with those of India in the knowledge that the proximity to India and her port of Trincomalee makes Sri Lanka's security of immense concern to India. On the other hand, Sri Lanka has taken every opportunity of strengthening her position through every counterpoise available to her.[53]

When the British left Sri Lanka in 1948 her main concern had been to secure herself against absorption by India. Jawaharlal Nehru's seemingly innocuous remark that Sri Lanka was 'an autonomous unit of the Indian federation' was misconstrued in Sri Lanka to imply that India had imperialistic designs on the island. Such an apprehension was given substance when K. M. Panikkar argued that Sri Lanka was a vital link in the defence strategy of India.[54] It is therefore not surprising that from 1948 to 1956 the United National Party (UNP) of Sri Lanka maintained a staunchly pro-Western stance.[55] On the eve of her independence, Sri Lanka concluded a 'defence pact' with Britain and allowed the stationing of a British naval base at Trincomalee and an air base at Katunayaka. At the same time, Sri Lanka emphasised her Commonwealth connections to ward off any possible Indian design.[56]

However, despite the continuing irritation of the Tamil problem, the Indo–Sri Lankan relationship improved considerably with the coming to power of SWRD Bandaranaike's Sri Lanka Freedom Party (SLEP) in 1956.[57] The easing of relationships was also facilitated by a thawing in the cold war: the process of de-Stalinisation had begun in the USSR and the influence of McCarthyism was beginning to wane in the USA. Bandaranaike therefore had no difficulty in persuading the British to evacuate their bases in Sri Lanka. Bandaranaike also shared with Nehru a belief in non-alignment and wanted an effective role for the small states in the community of nations. Although Sri Lanka refused to enter into a joint defence pact with India for fear of being drawn into India's orbit and getting entangled in India's conflicts, she nevertheless worked in concert with India at the UNO and accepted India's stand over Tibet in 1959. In 1961 Sri Lanka also supported wholeheartedly India's incorporation of Goa, a Portuguese enclave, and even banned the Portuguese ships on their way to Goa from entering her ports.[58]

Although there was a considerable unanimity between India and Sri Lanka on international questions, none of these impinged directly on either party. But the relationship was considerably strained during India's clash with China in 1962. Anxious not to be drawn into a conflict between two larger powers, Sri Lanka sought to maintain equidistance between India and China.[59] India's military débâcle and China's subsequent emergence as a regional power has enabled Sri Lanka to pull out of its dependence on India. Like Nepal, Sri Lanka has used India's discomfiture in the hands of the Chinese to assert a greater independence of action and decrease her economic reliance on India.

Although distance has precluded a particularly close relationship with Pakistan, Sri Lanka sees in Pakistan a possible ally in her constant quest for a counterpoise to India.[60] Pakistan has demonstrated that it can hold its own against India and therefore has an obvious attraction for Sri Lanka. During the 1970–71 crisis in Pakistan, Sri Lanka allowed over-flight by Pakistan International Airlines and turned a blind eye to the fact that troops were being carried to the then East Pakistan. It also refused to recognise Bangladesh for several months in deference to Pakistan's wishes.[61] However, Sri Lanka could not ignore, particularly in the aftermath of the 1971 Indo–Pakistan war, that India had emerged as the dominant power in the region and that, despite the US and Chinese backing, Pakistan was truncated. Prudence therefore demanded a closer relationship with India. Sri Lanka, together with India, has enthusiastically supported the idea of making the Indian Ocean a nuclear free zone.[62]

THE HIMALAYAN KINGDOMS

During the British rule the northern frontiers of India along the Himalayas were secured by a string of buffer states – Sikkim, Tibet, Bhutan and Nepal – perched up high on the mountainous slopes.[63] Britain had not found it necessary to incorporate these states into the Indian empire as she was able to secure her interest through special relationships with these Himalayan Kingdoms. Moreover, China which was on the other side of the mountains was a sleeping giant and posed no real threat to British India. India inherited the situation from Britain at independence and endeavoured to maintain a status quo in the lofty heights of the Himalayas. At first there was no cause for alarm. China had at last shaken off her opium-induced slumber, but was friendly enough and showed no immediate signs of interest in South Asia. There was no reason why the two great neighbours, separated by the Himalayas, could not exist back to back and pursue their own interests.

The geopolitical situation was dramatically altered when China moved into Tibet in 1950. At a stroke Tibet ceased to exist as a buffer and India's northern frontiers, despite the natural obstacles provided by the mountain ranges, lay exposed to penetration by China.[64] China's borders stretched closer to India; and Bhutan, Nepal and Sikkim now had common borders with China. India initially acquiesced in Chinese expansion and up until 1958 sought to safeguard her interest by the Sino–India Treaty of 1954 and by cultivating cordial diplomatic relations with China. But in 1959 when China crushed Tibet's autonomy and moved large numbers of Chinese troops to quell the rebellion, India was confronted with a large military power directly across her frontiers and the 1954 Treaty was rendered as a dead letter.[65] India's Himalayan strategy therefore acquired a compelling urgency. India strengthened and even redefined her relations with the three remaining buffer states – Bhutan, Nepal and Sikkim – and sought to integrate them more closely to India's defence system. But it was not an easy task as each of these Kingdoms had enjoyed a considerable degree of autonomy and even independence under the British and they were not willing to surrender their privileged position merely at the behest of New Delhi.

Nepal

Geography, history and culture have determined the relations of Nepal with India and to a lessen extent with China.[66] Geographically Nepal and India are one unit, and each can be regarded as the extension of the other depending whether one views it from the Himalayas or from the sea. Nepal's lowland areas are a part of the Gangetic Plain and it occupies the central part of the Himalayan foot-hills and mountains that extend for nearly 3000 miles between China and the sub-continent. Nepal skirts nearly 500 miles of India's northern border and to its west, south and east are the three Indian states of Uttar Pradesh, Bihar and Bengal. To the south-east it is separated from Bangladesh by 22 miles of Indian territory. To Nepal's north is Tibet. Although the Himalayas provide a natural boundary, there are several passes through which the Chinese can penetrate into Nepal. This became a particularly serious threat after China occupied Tibet, which until 1950 had been a buffer state between India and China. Consequently Nepal's position as a buffer is of great concern to India, especially as there are no natural frontiers between India and Nepal. India maintains an 'open' border with Nepal and should China or any other power occupy Nepal, the entire Indo–Gangetic plain would be exposed.

Ethnic and cultural links between Nepal and India are so close that they often blur the distinction between the foreign and domestic politics of the two countries.[67] Nepal is predominatly Hindu and over 25 per cent of its population has migrated from India during the present century. It has close linguistic affinity; Hindi is used quite widely and there are over a million Indian settlers in the fertile Terai region of Nepal. In India over 5 million Nepalese live and work and over a hundred thousand seasonal workers come to India every year. Nepali is now recognised as an 'associated official language' in West Bengal and the Indian Gurkha League is demanding the establishment of a Nepali-speaking state (province) in the Indian Union around Darjeeling.

Because Nepal is strategically essential to India, it has sought to develop what is euphemistically called a 'special relationship'.[68] While India recognises Nepal's independence, it will not brook any foreign interference in that Kingdom. India's interests in Nepal cover the entire gamut of political, military, economic and cultural relationships. The Indo–Nepalese Treaty of Peace and Friendship (1950) laid down that 'Neither Government shall tolerate any threat to the security of the other by a foreign aggressor.'[69] Nepal was also obliged to seek assistance from the Government of India before importing arms and equipment from a third country. Nepal must also give first preference to the Government of India or its nationals whenever it seeks assistance for the development of Nepal's natural resources or for the establishment of any industrial project.[70] It was also laid down that Nepal would not, without India's concurrence, employ any foreign (excluding UN) personnel whose activity may be prejudicial to the security of the two countries.[71] While both countries have often violated the treaty and the Nepalese Government has made claims that the treaty is obsolete, even if not abrogated, the Government of India has shown unwillingness to let the treaty default, and could invoke the treaty agreements if its interests in Nepal were seriously threatened.

But the relationship between the two countries has not always been harmonious and the Nepalese resent India's domineering attitude. In the early 1960s, the relationship 'cooled' when the new King, Mahendra, tried to balance Indian influence by seeking closer links with China and Pakistan.[72] The Sino–Indian war, by highlighting the strategic importance of Nepal, further weakened India's ability to twist Nepal's arm. The Indians were compelled to withdraw their military personnel from Nepal's northern frontiers and the military advisers from Katmandu. The relationship between the two countries reached a nadir in 1975 when India incorporated the autonomous state of Sikkim into India. Nepal, fearing a similar fate, began to distance itself from India by adjusting her relations with China and demanding that Nepal be declared a zone of peace.[73] But there are limits to what Nepal can accomplish given her near total dependence on India for trade and access to the sea.

Nepal is economically dependent on India: 60 per cent of Nepal's exports and 70 per cent of her imports are from India; India contributes 50 per cent of Nepal's foreign aid and provides oil, coal, petroleum products and cement at a highly subsidised rate. Nepal cannot replace these products at a comparable price from any other source.[74] Nepal is one of the poorest countries in the world and 90 per cent of her workers are engaged in agriculture. Her economy cannot cope with its growing population of 15 million. The 'open border' with India provides a safety valve. Some 5 million Nepalese have settled in India, many come for seasonal work and together they send home over 200 million rupees every year.[75] But by far the greatest leverage which India has over the landlocked Nepal is the transit facility to the outside world.[76] At present almost all exports and imports in Nepal from outside the sub-continent come via the port of Calcutta and are then freighted overland across Indian territory into Nepal. The transit arrangements are carefully defined in trade and transit treaties which are periodically renewed. Nepal has made numerous representations that transit facilities for landlocked countries should be an internationally guaranteed right but India has turned a deaf ear. In 1976, when relations were strained between the two countries, India stalled the renewal of the treaty for nearly two years. It was a sufficient signal for Nepal to tone down its hostility to India. As relations improved, the two treaties regulating transit and trade were concluded in 1978, and this gave Nepal a favourable deal and met several of its demands: increased number of transit points, more warehouse spaces and port facilities, simpler customs procedures, and removal of many restrictions in Nepal's re-export to India.[77] Some of the trade regulations are likely to be further eased as India itself is relaxing its export–import facilities. But so far India has not agreed to allowing Nepal to use the port of Chittagong in Bangladesh. In many ways this would make good sense. The port of Chittagong is modern, not congested and well served by railways right up to the northern tip of Bangladesh – which is separated from Nepal by 22 miles of Indian territory. It is conceivable that this alternative transit route may soon become available as it would also enormously benefit India's own movement of goods from Calcutta to Assam, northern Bengal and even Bhutan. But this would depend on India's relationship with Bangladesh improving – again a distinct possibility should a democratic regime be restored in Bangladesh. In recent years Nepal's communication has greatly improved. It is now possible to travel from East to West in Nepal without passing through Indian territory; there is a direct all-weather road to the Tibetan border; but as yet travel between Nepal and Bhutan is only possible through India.

Although Indo–Nepalese relations are better than they were in 1975, they are still far from cordial. There is a considerable apprehension that when the Trade and Transit Treaties come up for renewal later this year (1985), India may not be as generous as in 1978. India is irritated by Nepal's dilatoriness in developing the hydroelectric protects in Nepal which is vital for the power-starved industries of West Bengal, Bihar and Uttar Pradesh. More recently the removal of the pro-Indian Prime Minister S. B. Thopa by L. B.

Chand has not only introduced domestic instability in Nepal but further strained the relations with India. In recent years one of the main causes for India's annoyance with Nepal is King Birendra's proposal to have Nepal declared as a 'zone of peace' with international guarantees. India, of course, suspects this to be a ploy to undo the Indo–Nepalese treaty of 1950 and a subtle move to put India and China on equal footing in Nepal. India's response was predictably unenthusiastic because the proposal sought to take Nepal outside India's defence perimeter. According to India the necessity to declare Nepal' 'a zone of peace' was superfluous because Nepal's security was guaranteed by the Indo–Nepalese Treaty of 1950. But India's main objection is that it ignores the geopolitical realities: whereas India's presence in Nepal does not threaten China, the converse is not true. But it seems doubtful if India can resist the plan should Nepal find sufficient number of international sponsors.

While Nepal's dependence on India is substantial, India's leverage over her is limited by strategic considerations. Moreover, as India and Nepal develop their joint economic ventures, particularly hydroelectric, the economic relationship will become a two-way traffic. While India will resist any foreign power displacing her 'special position' in Nepal, she is unlikely to obstruct Nepal from developing economic contacts with foreign companies which are of benefit to Nepal. Besides in the wake of the Soviet occupation of Afghanistan and the turmoil in Kampuchea and Northern Burma, Nepal is anxious not to upset equilibrium in the region: it would stand to lose either if China gained hegemony in the Himalayas or if India was weakened by Sino–Pakistan collusion. Non-alignment and friendship with India is Nepal's strategy for survival.

Bhutan

Bhutan, like Nepal, is of immense strategic importance to India because of its geographical location. Situated between India in the South and China in the north, it is pulled by both great neighbours.[78] Bhutan's ethnic and religious links have historically been mainly with Tibet, which is also her closest neighbour. But with the Chinese occupation of Tibet and the closure of its borders, Bhutan turned more to its next closest neighbour India with whom she has special treaty relations.[79]

Indo–Bhutanese relationship is defined by a treaty of 1949 in which India acknowledges the sovereign independence of Bhutan but at the same time, perhaps slightly paradoxically, 'the government of Bhutan agrees to be guided by the advice of the Government of India in regard to its external relations'.[80] Tiny, lacking in resources, landlocked and depending on neighbours for access to the outside world, Bhutan has had no option but to acquiesce in her somewhat ambivalent status. Bhutan's precariousness was increased by the fate of Tibet, which brought China nearer to her than ever before.

After the Sino–India war of 1962, Bhutan, in marked contrast to Nepal, moved closer to India and is now heavily dependent on the latter both politically and economically.[81] India supplies Bhutan with much of her aid requirements, has built roads and trained the security forces. But India has not taken Bhutan's dependence for granted and has been careful not to alienate Bhutanese sentiment.[82] In the period since 1962 India has recognised Bhutan's ability to play China against her and has therefore handled Indo–Bhutanese relations with considerable finesse. In 1961 Bhutan was allowed to join the Colombo Plan; in 1971 India sponsored Bhutan's membership of the UNO; Bhutan established its first foreign legation in New Delhi in 1978 and followed it shortly afterwards with an embassy in Bangladesh. Bhutan also joined the non-aligned movement in 1979. So far Bhutan has refrained from following Nepal's example of establishing closer relations with China as that could be a potential source of irritation between India and Bhutan. For the moment the relationship is mutually convenient and each is aware of its own limitations. The memory of Tibet and the experience of Sikkim points Bhutan to the expedience of good relations with India; the ability of Nepal to counter-balance Indian influence through Chinese involvement is a sharp reminder to India not to become too imperious or overbearing.

Sikkim

Sikkim, sandwiched between Bhutan and Nepal, shares a common frontier with Tibet in the north, and thus provides a possible wedge for China's penetration into India's north-eastern region. Sikkim's strategic importance, like that of Nepal and Bhutan, was highlighted during India's clash with China in the North-Eastern Frontier Agency. The control of Sikkim by an unfriendly power would be a serious menace and make the defence of India's north-eastern perimeter almost an impossible task.

Sikkim was not an independent state like Nepal or an independent state under protection like Bhutan but a protectorate of India until its incorporation as the twenty-second state of the Union of India in 1975.[83] Sikkim's relations with British India were defined by the Treaties of 1861 and 1918 whereby she was accorded the status of a princely state. But the independent Indian government did not integrate Sikkim on the same basis as other princely states but instead concluded a separate treaty in 1950. This treaty accorded Sikkim the status of a protectorate with full internal autonomy but reserved for the Government of India the control of foreign policy and the right to defend Sikkim. Autonomy in internal affairs was constrained by India's ultimate responsibility for good government.[84]

India was content to maintain the status quo in Sikkim so long as the Chogyal rulers accepted their ceremonial role and harboured no pretensions to conduct their own foreign relations. But when in 1973 the Chogyal became restless and evinced a desire for increased international status, India invoked her responsibility for good government and thrust a 'popular government' on Sikkim. But as the Chogyal would not easily submit to New Delhi's dictates, Sikkim was finally incorporated into the Union of India as a state. India thus demonstrated her determination not to allow a weakening of her defence system in the north for which the buffer states are essential. It also showed how little manœuvrability is available to the small states in the face of India's resolution.

AFGHANISTAN

Afghanistan which impinges heavily on Pakistan's domestic security has already been dealt with in an earlier chapter. However, historically, Afghanistan's role has been that of a buffer between India's north-west frontiers and Russia.[85] The British Indian strategy was obsessed with keeping away the Russian Empire (and after 1917 the Soviet Union) from what it perceived to be a determined attempt to expand to the warm waters of the Indian Ocean and to the oil wells of the Gulf. Although the Soviet penetration of Afghanistan had begun immediately after the British evacuated the sub-continent, Afghanistan nevertheless served its purpose by keeping direct Soviet presence away from Pakistan's borders. However, the situation was dramatically altered by the Soviet occupation of Afghanistan in the last days of 1979, and has introduced a new element in the security perceptions of Pakistan and also to those of India.

Pakistan is now faced with hostile powers on both her frontiers, extending for over a thousand miles on each side and stretching her already extended flanks. While the presence of the Soviets in Afghanistan, together with three million Afghan refugees in Pakistan have added to the country's security anxieties, curiously enough Pakistan is still convinced that the real threat comes from India. The bulk of Pakistan's army, thirteen infantry divisions and two armoured divisions, are deployed to guard the eastern frontiers.

Pakistan's security analysts, obsessed with the threat from India, maintain that despite the Soviet annoyance with Pakistan for allowing her territory to be used by the Afghan Mujahideen to fight against the pro-Soviet government in Kabul, the Soviet Union would not be willing to risk a direct confrontation with Pakistan.[86] A Soviet intervention in Pakistan, it is argued, would provoke China and/or the USA to come to Pakistan's aid. Nor could a Soviet-backed Afghan invasion of Pakistan be easily mounted because the existing Soviet troops in Afghanistan are already pinned down fighting the guerrillas. For the Afghans to take on Pakistan would call for massive reinforcements from the Soviet Union which even if it were logistically possible would take a long time to complete and give plenty of warning to Pakistan's allies to come to her aid. Rather than confront Pakistan head on, according to this scenario, the Soviets might encourage India to move against Pakistan. It is further argued that India, which has close links with the Soviet Union and under considerable obligation for military and economic aid, might be willing to act as proxy for her Soviet allies. Moreover, such a policy would dovetail with her own preferences in Afghanistan. The Indian government would prefer to see a Marxist government in Kabul to a fundamentalist Islamic regime and suspects the resistance leaders of being largely pro-Pakistani. Moreover, India can hope to exert little diplomatic pressures on the Soviet Union given the importance the Soviets attach to Afghanistan.

Obviously a Soviet-inspired Indian attack on Pakistan would be logistically much easier and would raise comparatively fewer international complications. Most of India's troops and armour are already deployed along the Pakistani frontiers and therefore could be easily mobilised. Moreover, such a conflict would be seen as one more instance of the South Asian imbroglio in which the outside powers would be unwilling to be drawn in. From India's point of view such a war would have certain advantages: it would give India the opportunity of destroying Pakistan's military might before she actually acquires a nuclear bomb or builds up conventional arms superiority through US assistance; and second, Indian forces by pressing north for a link-up with the Soviet forces in Afghanistan would not only capture the Pakistani part of Kashmir but also cut off Pakistan's overland supply route from China. If such a scenario were to materialise, the Soviets in return for using their influence on India for desisting from invasion would extract a promise from Pakistan that it would withdraw its support and bases from the Afghan rebels.[87]

While the invasion scenario sounds plausible, and might even make sense in terms of game theory, it does not take into account fully the possible reactions of the USA and China. The scenario, in fact, reflects a Pakistani obsession with India and as one suspects, a useful ploy for securing more arms from her allies. This also underestimates India's own security considerations. Even though the USSR is a close ally, India cannot view the developments in Afghanistan with equanimity. India considers the existence of Afghanistan to be a buffer essential to the security of South Asia and is therefore anxious to see a phased withdrawal of the Soviet forces.[88] New Delhi is apprehensive that a consolidation of the Soviet power in Afghanistan might eventually lead to an encroachment on, and eventual disintegration of Pakistan. While India would have some reasons for satisfaction to see the splintering of Pakistan, it is equally wary of allowing a direct Soviet involvement in the sub-continent which India regards as her stamping ground. The Soviet intrusion, India fears, might recreate the sort of problems created by China's involvement in South Asia. Behind the scenes India has brought pressures on the Soviets to evacuate Afghanistan and understandably it is critical of Pakistan and the USA which it sees as hindering the chances of the Soviet withdrawal by destabilising the Babrak regime.[89] While a conflict with Pakistan under Indira Gandhi may have been a possibility, Rajiv Gandhi seems more committed to a peaceful sub-continent. Besides, with his hands already full with troubles in the Punjab, it is highly unlikely that he would opt for confrontation with Pakistan.

Pakistan itself has been pursuing a conciliatory policy and has launched a three-pronged approach to reduce risks arising from the Afghan crisis. First, it has adopted a cautious policy towards the USSR applying gentle diplomatic pressures in negotiation but careful to avoid any direct provocation. It has offered the Afghan refugees sanctuary but has refused to supply arms directly to the rebels and there is no evidence of any collusion between the Mujahideen and the Pakistan army. Second, Pakistan has offered India a 'no-war pact' arguing that the presence of the Soviet Union has altered the geopolitical situation in the region. In the recent diplomatic exchanges between India and Pakistan, the possibility of some such agreement appears brighter than ever before. Pakistan has been careful not to provoke India by getting entangled with the Sikh secessionists. And finally, Pakistan has made skilful use of the Afghan crisis to extract a massive 3.2 billion dollar military and economic package from the USA to improve her defence capability. Pakistan obtained the aid without compromising either her non-aligned status or her links with China and the Islamic world.

CONCLUSION

The smaller states of South Asia because of their relatively small size, population, poverty and geographical location, cannot hope to exert much influence in the South Asian complex. The lopsided bipolar structure of South Asia remains intact. None of the smaller states have any direct alliance with Pakistan: geographical proximity to India and her pre-eminent position makes such alliances politically undesirable if not actually impossible. On the other hand, the emergence of Bangladesh has neither weakened Pakistan nor substantially altered the balance in India's favour. In fact, it may have created more problems for India and made the problems of handling the small states more tricky. Bangladesh with its hundred million people defies any possibility of being absorbed by India. By emphasising her links with Pakistan and the Islamic world, Bangladesh has substantially reduced her dependence on India – even if not always to her advantage – and exercises greater freedom of action than her impoverished status could beget. Further, by advocating multilateral diplomacy through a regional forum for South Asia it has sought to reduce India's predominance inherent in bilateral relationships between large and small states.

India's ability to exert too much influence on the smaller states is limited by two factors. First, while India is relatively better off economically and has provided considerable aid and assistance to her smaller neighbours, she is by no means in any position to satisfy all their development needs. The smaller states have therefore cast their net widely and sought aid from different donors. Second, the smaller states have taken advantage of the Indo–Pakistan and Sino–Indian rivalry to enhance their independent status and extract important economic and political concessions which given the power imbalance these tiny states could scarcely have contemplated. But this must not be stressed too far. In the case of Nepal and Bhutan, India can also exert considerable leverage over them because, apart from supplying their essential commodities like oil and manufactures, it also virtually controls their access to the outside world.

The traditional role of some of the smaller states as buffers is still important. India continued the British strategy of providing depth for defence in her north and north-eastern frontiers through a series of buffer states: Afghanistan, Tibet, Bhutan, Sikkim and Nepal. In the early years of independence India maintained her 'special relationship' with Nepal, Bhutan and Sikkim whereby these states enjoyed varying degrees of internal autonomy but their foreign relations were restricted so that they did not in any way hinder India's defence strategy. However, the continuation of the co-called 'special relationship' was rendered difficult by two important developments. First, the occupation of Tibet by China brought the Chinese right up to India's doorstep and lay open her frontiers to Chinese penetration. While this dramatically exposed India's vulnerability and emphasised the importance of the other buffer states for India's security, the continuance of the 'special relationship' was further complicated by the Sino–Indian war of 1962. The fear of alienating China compelled Nepal to distance itself from India and, by successfully playing India and China against each other, has established greater freedom of action both in domestic and foreign affairs. Bhutan, on the other hand, frightened by the fate of Tibet, moved closer to India. India too, mindful of its experience with Nepal, has shown greater consideration to Bhutan than in the past. To an extent an international recognition of Nepal and Bhutan's independence fits in well with India's strategy. By emphasising their independent status and the membership of the UNO, India hopes to prevent their absorption by China. Sikkim, which albeit had the status of a protectorate, was less fortunate in trying to assert its autonomy by using the China card and ended up by being incorporated into the Union of India. It is clear that India attaches the highest importance to the buffer states and would certainly resist the presence of any foreign power in these states which is detrimental to her security.

The position of Afghanistan, which traditionally insulated South Asia from the Russians, remains uncertain. The occupation of the country by the Soviet Union threatens to end its role as a buffer. The prospect of the Soviet withdrawal from Afghanistan in the near future seems doubtful, and the full implication of this for the South Asian security complex is difficult to assess. It could be the beginning of a direct Soviet involvement in South Asia with far reaching consequences for the region.

Neither the USA nor the USSR (despite its invasion of Afghanistan) have as yet any direct interest in South Asia; their involvement in the region is born of super-power rivalries and global geopolitical considerations. (Even China's interest in the region is necessarily limited to her border conflict with India.) Consequently the smaller states with varying degrees of success have managed to draw bigger powers into South Asia and have thereby successfully checked what they have perceived to be India's hegemonic pretensions. Like Pakistan, but on a more modest scale, Bangladesh, Sri Lanka and Nepal have all sought foreign assistance to offset India's influence. The price of foreign assistance, however, has not always been cheap. Between 1948 and 1956, Sri Lanka strengthened its military position through an alliance with Britain. But what it gained from India in terms of freedom of action, it lost to Britain by having to concede air and naval bases in her territory. Alliance with foreign powers has given the smaller states a counterpoise to reduce Indian domination. But in the long run they are jumping from the frying-pan into the fire. None of the smaller states can shake off the impediment of resources, size and location which nature has thrust on them.

NOTES

1. B. H. Farmer, *An Introduction to South Asia* (London, 1983) ch. 1.
2. W. J. Barnds, 'South Asia', in J. W. Rosenau *et al.*, *World Politics* (New York, 1976) pp. 501–27.
3. See Ramakant and M. D. Dharamdasani, 'Pakistan–Nepal Relations; from Indifference to Consolidation', pp. 426–36, and N. Iyer, 'Pakistan and Sri Lanka: the Dynamics of Distant Cordiality', pp. 437–53, in S. Chopra (ed.), *Perspectives on Pakistan's Foreign Policy* (Amritsar, 1983).
4. Z. Khalilzad, *Security in Southern Asia, I: The Security of Southwest Asia* (Aldershot, 1984) p. 99.

5. S. Mansingh, *India's Search for Power* (New Delhi, 1984) ch. 6.
6. Urmila Phadnis, 'Indo–Ceylonese Relations', in M. S. Rajan (ed.) *India's Foreign Relations During the Nehru Era* (Bombay, 1974) pp. 26–31.
7. T. Maniruzzaman, 'Bangladesh in 1976: Struggle for Survival as an Independent State', *Asian Survey*, vol. XVII, no. 2 (February 1977).
8. K. Subramanyam, 'India's Security – Present and Future', paper presented at a Seminar on 'India's security: Past, Present and Future' at St Antony's College, Oxford, October 27–29, 1984.
9. Ibid.
10. D. M. Prasad, 'Indo–Sri Lanka Relations; Mutual Problems and Common Approaches', in S. Chopra (ed.), *Studies in India's Foreign Policy* (Amritsar, 1983) pp. 297–8.
11. *Bangladesh Observer*, 10 October 1984.
12. K. Subramanyam, 'India's Security – Present and Future'.
13. For details, see A. Toussaint, *History of the Indian Ocean* (Chicago, 1969).
14. T. George, R. Litwak and Shahram Chubin, *Security in Southern Asia 2: India and the Great Powers* (Aldershot, 1984) pp. 116–23.
15. Ibid., pp. 2–8.
16. R. Rahul, *Royal Bhutan* (New Delhi, 1983) pp. 61–2: R. and M. D. Dharamdasani, 'India's Attitude Towards Nepal', in Chopra (ed.), *Studies in India's Foreign Policy*, pp. 281–94.
17. Barnds, 'South Asia', pp. 508–13.
18. See R. Sobhan, *The Crisis of External Dependence. The Political Economy of Foreign Aid to Bangladesh* (Dacca, 1982).
19. R. Jahan, 'Bangladesh Nationalism', in E. Ahamed (ed.), *Bangladesh Politics: Problems and Issues* (Dacca, 1980) pp. 93–127.
20. M. Anisuzzaman, 'Bangladesh Nationalism', in E. Ahamed (ed.), *Bangladesh Politics* (Dacca, 1980) pp. 79–97.
21. I. Hossain, 'Bangladesh–India Relations', in E. Ahamed (ed.), *Foreign Policy of Bangladesh: A Small State's Imperative* (Dacca, 1984) p. 36.
22. Jahan, *Bangladesh Politics*, pp. 205–206.
23. Sobhan, *The Crisis of External Dependence*, pp. 139–43.
24. For details, see S. S. Bindra, *Indo–Bangladesh Relations* (New Delhi, 1982).
25. S. M. Ali, *After the Dark Night: Problems of Sheikh Mujibur Rahman* (New Delhi, 1977) pp. 136–7.
26. I. Hossain, 'Bangladesh–India Relations'.
27. *Amrita Bazar Patrika* (Calcutta), 27 November 1977; *The Statesman* (Calcutta) 20 and 21 December 1977.
28. R. D. Sawvell, 'Crisis on the Ganges: the Barrage at Farakka', *Geography*, Vol. 63 (1978) pp. 49–52; for a Bangladesh point of view, see B. M. Abbas, *The Ganges Water Dispute* (Dacca, 1983).
29. *Asian Recorder* (4–10 June 1974) pp. 12 and 37.
30. Mansingh, *India's Search for Power*, p. 298.
31. S. A. Hussain, 'Security Perceptions of Bangladesh: Dilemmas of a Small State', a paper in a seminar on 'South Asian Security: Past, Present and Future' held at St Antony's College, Oxford from 27 to 29 October 1984.

32. *Far Eastern Economic Review*, 7 December 1979.
33. Ibid., 7 December 1979.
34. Hussain, 'Security Perceptions of Bangladesh', pp. 10–11.
35. *The Statesman*, 6 July 1980.
36. *Overseas Hindustan Times*, 24 January 1980.
37. I. Hussain, 'Bangladesh–United States Relations: The First Decade', in E. Ahamed (ed.), *Foreign Policy of Bangladesh*, pp. 64–80.
38. S. D. Muni, 'Strategic Aspects of SARC', *Strategic Analysis,* vol. IX, no. 4 (April 1984), pp. 24–5.
39. See S. U. Kodikara, *Indo–Ceylon Relations Since Independence* (Colombo, 1965).
40. Ibid., p. 164.
41. Phadnis, 'Indo–Ceylonese Relations', pp. 27–9.
42. Ibid., p. 29.
43. P. Mason (ed.), *India and Ceylon: Unity and Diversity, A Symposium* (London, 1967) p. 274: see also W. H. Wriggins, 'Impediments to Unity in New Nations: The Case of Ceylon', *American Political Science Review*, vol. LV (June 1961).
44. R. B. Goldmann and A. Jeyaratnam Wilson (eds), *From Independence to Statehood: Managing Ethnic Conflict in Five African and Asian Countries* (London, 1984). See esp. 7, 8, 9 and 11.
45. 'Brink of Civil War', *Far Eastern Economic Review* (21 February 1985) pp. 36–41.
46. Ian Jack, 'Sri Lanka's Last Bid to Woo Tamils from Terror', *The Sunday Times*, 16 December 1984; The *Guardian*, 9 May 1985.
47. 'Brink of Civil War', *Far Eastern Economic Review* (21 February 1985), pp. 36–41.
48. The Sunday *Observer*, 3 March 1985.
49. Jack, 'Sri Lanka's Last Bid to Woo Tamils from Terror.'
50. 'Hope on the Horizon', *India Today* (30 June 1985) pp. 60–61.
51. *India Today* (15 February 1985).
52. Prasad, 'Indo–Sri Lanka Relations', pp. 296–7.
53. S. D. Muni and Urmila Phadnis, 'Ceylon, Nepal and the Emergence of Bangla Desh', *Economic and Political Weekly*, vol. VII (19 February 1972), pp. 471–6.
54. K. M. Panikkar, *The Strategic Problems of the Indian Ocean* (Allahabad, 1944), pp. 16–18.
55. K. P. Krishna Shetty, 'Ceylon's Foreign Policy: Emerging Patterns of Nonalignment', *South Asian Studies*, vol. I (April 1966) pp. 3–14.
56. Prasad, 'Indo–Sri Lanka Relations', p. 297.
57. Phadnis, 'Indo–Ceylonese Relations', pp. 23–5.
58. *Asian Recorder*, vol. VIII (15–21 January 1962) p. 4370.
59. Urmila Phadnis, 'Ceylon and the Sino–Indian Border Conflict', *Asian Survey*, vol. III, pp. 189–96.
60. Prasad, 'Indo–Sri Lanka Relations', p. 298.
61. Muni and Phadnis, 'Ceylon, Nepal and Bangladesh'.
62. C. Kumar, 'The Indian Ocean: The Arc of Crisis or Zone of Peace', *International Affairs* (Spring 1984) pp. 234–46.
63. By far the best work on Nepal is Leo Rose, *Nepal: Strategy for Survival*

156 *The Role of the Smaller State in the South Asian Complex*

(London, 1971); see also P. C. Chakravarti, *The Evolution of India's Northern Border* (New York, 1971).
64. For details see P. P. Karan and W. M. Jenkins, *Nepal: A Cultural and Physical Geography* (Lexington, 1960).
65. S. Kumar, 'India and Nepal', in M. S. Rajan (ed.) *India's Foreign Relations During the Nehru Era* pp. 63–5.
66. R. and M. D. Dharamdasani, 'India's Attitude Towards Nepal', pp. 281–2.
67. N. Kihara (ed.), *People of Nepal Himalayas* (Koyota, 1957) *passim*.
68. R. Shaha, *Nepali Politics: Retrospect and Prospect* (New Delhi, 1976) pp. 142–4.
69. A. S. Bhasin, *Documents on Nepal's Relations with India* (New Delhi, n.d.) p. 23.
70. Full text of the letter exchanged at the time of the signing of the Treaty is reproduced in S. D. Muni, *Foreign Policy of Nepal* (Delhi, 1973).
71. Ibid.
72. Shaha, *Nepali Politics*, pp. 144–7.
73. R. and M. D. Dharamdasani, 'India's Attitude Towards Nepal', pp. 291–2.
74. Shaha, *Nepali Politics*, pp. 158–64.
75. S. D. Muni, *Foreign Policy of Nepal*, (Delhi, 1973) pp. 35–46.
76. Ibid., pp. 77–78.
77. Mansingh, *India's Search for Power*, pp. 283–88.
78. R. Rahul, *The Himalaya as a Frontier* (New Delhi, 1978) p. 96; S. Mansingh; *India's Search for Power*, pp. 296–7.
79. M. Kohli, 'The China Factor in Indo–Bhutanese Relations', in S. Chopra (ed.), *Studies in India's Foreign Policy'*, p. 163.
80. Mansingh, *India's Search for Power*, pp. 296–7.
81. J. Belfiglio, 'India's Economic and Political Relations with Bhutan', *Asian Survey* (August 1972).
82. M. Kohli, 'Dragon Kingdom's Urge for an International Role', *India Quarterly*, vol. xxxvii, no. 2 (April–June 1981) pp. 227–40.
83. Mansingh, *India's Search for Power*, pp. 281–3.
84. Ibid.
85. D. P. Singh, *India and Afghanistan 1876–1907* (Delhi, 1963).
86. Z. Khalilzad, *Security in Southern Asia 1. The Security of Southwest Asia* (Aldershot, 1984) p. 106.
87. Ibid. p. 107.
88. B. Sen Gupta, 'The Necessity of Choice', *Seminar* (February 1980), p. 36.
89. See A. B. Vajpayee's statement in *Times of India*, 2 January 1980.

Part IV
The Super-regional Component of the Security Problem

From the discussions in the previous two parts, it becomes clear that Indo–Pakistan rivalry, despite the possible easing of tension in Kashmir, is an enduring one. This inevitably means that Pakistan, in order to maintain a balance of power in the sub-continent, must endeavour to enhance its alliances with outside powers in order to offset India's natural pre-eminence. Pakistan is fortunate in the sense that it can play its Islamic card both to draw in the support of the Gulf states and also to cement its relationship with Bangladesh.

While none of the smaller states of South Asia have a particularly close link with Pakistan, Pakistan has given them greater leeway in their relationships with India by challenging India's hegemony. Despite India's forcible absorption of Sikkim and Goa, its ability to pursue a heavy-handed policy with its smaller neighbours is limited by their ability to secure the assistance and support of outside powers. Thus Nepal has discarded India's claim to a 'special' relationship by tilting towards China; Bangladesh by emphasising its Islamic links, and by various overtures to the United States, has moved away from India; and Bhutan, no doubt worried by the fate of Sikkim and Tibet, has enhanced its independent status by exploiting the Sino–Indian rivalry. Only Sri Lanka, because of its proximity to India and its great distance from any other countervailing power, has not been particularly successful in offsetting India. Sri Lanka has had to be contented with strengthening its position through membership of the Nonaligned Movement and other international bodies. The presence of a large number of Tamils in the northeast, now virtually in a state of civil war, gives India an additional leverage in the affairs of Sri Lanka. But Sri

Lanka, with its excellent port at Trincomalee overlooking the Bay of Bengal, has obvious attractions to the superpowers engaged in the Indian Ocean should it wish to seek outside support.

In this part, our focus is on the security interactions between different complexes. For reasons set out in the Introduction, we look at the lateral relationship between South Asia and the Gulf/Middle East, and at the hierarchical relationship between South Asia and the Sino–Soviet complex, but not at the lateral relationship between South Asia and Southeast Asia. The theoretical proposition that lateral relations are characterised by relative indifference, and hierarchical ones by intervention from higher to lower, will also be examined.

In the case of the Gulf/Middle East, our attention is drawn to the possibility that the relative indifference between the security dynamics of the two complexes may be breaking down. We examine the significance of the military ties between Pakistan and the Gulf states, and ask whether they add materially to Pakistan's attempt to balance India. We ask also whether these ties reflect an engagement by Pakistan and the Gulf states in each other's rivalries, or merely a detached form of self-interested power-aggregation without deep political significance. We also begin to assess the impact of the escalation in superpower engagement in the boundary area between the two local complexes since 1979.

In the case of the Sino–Soviet complex, and the emergent Asian supercomplex that it seems to be generating, our principal interest is in the character and extent of the penetration from higher complex to lower, and the impact of it on the security of the South Asian states. We examine whether Sino–Soviet penetration of South Asia occurred more as an offshoot of rivalries among the states of the sub-continent, or more as a consequence of Sino–Indian disputes, and how much it was conditioned by the prior Soviet–American penetration. In particular, we look at the part played by the local pattern of insecurity in facilitating or resisting the penetration, and at how important South Asia is to the Sino–Soviet rivalry. The answers to these questions enable us to assess how deep and durable are Chinese and Soviet interests in sustaining their current patterns of involvement. Finally, we investigate whether the impact of the Sino–Soviet penetration on the security dynamic of the local complex has been to reinforce the local structure by intensifying its insecurities, or ameliorate it by imposing stability. We ask also whether the impact of the higher-level intervention has favoured one side or the other in terms of the distribution of power.

6 South Asia and the Gulf Complex

B. A. ROBERSON

INTRODUCTION

In an uncertain world, states search for the means to ensure their survival. This chapter will argue that the linkages that are developing between the Gulf states, in particular Saudi Arabia, and Pakistan may be leading to each acquiring obligations to the other's security as a means of contributing to the maintenance of their own status quo. The study will consider the scope and extent of these linkages. The chapter will conclude with an assessment of the ties that exist between Pakistan and the Gulf with a view to determining the durability and significance of the security relationships that have been formed. The focus will be largely on military issues.

Overwhelmingly, the literature on security has concentrated on the interests and needs of the West or the Soviet Union. Recent considerations of the states in the Third World have begun to focus more sharply on their particular problems of security, quite apart from the usual linkage to great power security interests.[1] It has become apparent that states in the Third World have been handling the problems and crises in their regions in a manner that limits the intervention of the superpowers. This is a prime characteristic of non-alignment which, however, did not preclude the search by Third World states for ways to use the superpowers' global interests in the service of their own needs.

The littoral states of the Persian Gulf have strong ties in their Islamic beliefs and, save for Iran, Arab heritage. In South Asia, Pakistan was founded as an Islamic state and as such is considered an integral part of the Islamic world. All these states have basic problems of domestic stability caused in part by tribalism and/or factions which are based on differing social and political goals, and sectarian cleavages.

Distinguishing between the processes of the nation, and those of the state, is one way of enhancing an understanding not only of Saudi Arabia but of the other Middle Eastern/South Asian states and their inter-state relations.[2] It is well to remember that there is a cultural community in the Middle East as a whole, in the form of the Umma (the ideal Islamic community). This community, an important general source of identification, is composed of numerous ethno-cultural societies each with enough historical variation to contribute to a discernible differentiation among Arab-speaking peoples. The relationship between this over-arching community and the local ethno-cultural societies that pepper the Middle East informs the continually evolving life of the area. The development of states' institutional frameworks has come rather late in the life of the Community and reflects, on the whole, the colonial legacy. These state structures have yet to catch up with the deeper, and constantly evolving, cultural and political experience of the peoples in the area. In each of the Middle Eastern countries there has been created a *de facto* regime with political power which is lacking in legitimacy and is in tension with religion and the larger community. The development of the national and cultural community both at the over-arching and local levels has not coincided with the state structures introduced into the region by external powers.

A key element in the extension of the practices of the European state system to the Middle East has been the fixing of identifiable boundaries in the period after the First World War. The location of these boundaries was not, on the whole, the result of indigenous political processes and as such contributes to their artificiality. *De facto* sovereignty over the people and territory within these boundaries occurred for the most part after 1945. The tension that exists between the larger community and the state is further exacerbated in the Islamic concept of the state. This neither endows the state with the right to claim the allegiance of the individual over all other things, nor accepts the Western concept of sovereignty which makes the state the preeminent entity in both domestic affairs and the international system.[3] God is ... 'the only sovereign ... [and] no Islamic state can claim absolute sovereignty' in the Western sense of the term.[4]

Because these states are new, the pattern of their relationships is in an early stage of development. A prominent feature of the developing Middle East state system is a constant pattern of realignment which has thus far prevented the rise of a hegemonic power. This feature underlines the supreme difficulty in establishing regional security, a feature within which the Arab statesmen move and breathe. The relationship between these states is intense and viewing them over the post-Second World War period it is clear that relationships shift and change according to perceived interests which reflects a constant state of instability.

THE GULF COMPLEX

Looking at the Gulf over several centuries, one can see that the development of the historical, ethno-cultural and political society differs markedly from the rest of the Middle East. The tribes of the central area of the Arabian Peninsula came under the rule of the Al Saud, a tribe of the Anazah confederation, who had formed an alliance in the mid-1700s with the Al-Wahhab family, founders of a militant religious movement whose purpose was to cleanse Islam of corrupt practices.[5] It was a movement imbued with a potent ideology which was both political and revivalist in the millennarian sense[6] and became the glue holding together the tribal organisation of the Peninsula and underwriting the centralised authority established by Abdel Aziz Al-Saud. The result was the Kingdom of Saudi Arabia in 1932 which in its constitution harks back to its early Islamic forebear of the seventh century. To a certain extent, Saudi Arabia as a newly formed independent state dealt successfully with the problem of dissonance between state formation, which is recent, and the much older national and cultural community. In addition, Saudi Arabia is distinct in one important respect. Due, perhaps, to its inhospitable geography, it has never experienced colonialism.[7] Saudi Arabia developed its own institutions and structures, in its own Arab way, and has no post-colonial inhibitions in dealing with the Great Powers of the day.

Saudi Arabia, a major power in the region in the 1930s, entered the post-Second World War era as a second rate power. This was a result of the changing forces in the region that occurred with the creation of Israel and the growing requirements of military capability. The Saudi response to the changed circumstances was to begin the modernisation of its military forces. But this process was constrained by both the slender means available to the government at the time, and the need to find forms of modernisation which would not pose threats to political control by the Saudi royal family.

The point to bear in mind about Saudi security in the post-1945 period is the royal family's concern to neutralise any threats to its rule arising from dissension either within the family or among the tribes, or from the military. This concern is reflected in the structure of the Saudi military – its division into three unintegrated commands (Ministry of Defence and Aviation, Ministry of Interior, and Presidency of the Council of Ministers) comprising five distinct and unco-ordinated military units (Army, Air Force, Navy, Frontier Forces and Coast Guard, and the National Guard[8] – and the manner in which modernisation of the military has been approached. Diversification of sources of military supplies have added to the complication of the management and organisation of the military. A more fundamental problem for the modernisation effort is the smallness of the manpower base and the illiteracy of the military recruits.[9]

There is a similar problem for the development and modernisation of the economy. The specific security puzzle of the Saudi government is the vast size of the country in which there are perceived threats at every corner. The necessity of deploying armed forces in sufficient strength at all vital points exposes their essential weakness – that there are not enough native Saudis to supply the requirements of Saudi military security. This essential vulnerability perhaps explains the fact that the government rarely takes censuses, and when it does, seems to accept the results with great reluctance. Estimates of the population range from 4 million to 10 million, but it is probably closer to 5 million.[10] Not only is the size of the population small, but also the illiteracy rate in the early 1980s was around 85 per cent.[11] Both factors place serious constraints on development programmes and security prospects. Just as the economic development programmes have had to depend upon large numbers of foreigners to carry them out (estimated at about 1.6 million), so also security policy has required not only foreign military officers, advisers, and trainers, but also foreign troops, especially since the 1970s. Saudi Arabia is dependent on outside assistance for dealing with a major threat and, depending on circumstances, would not be able to handle attacks on two fronts at the same time.

Similar security problems are discernible in the smaller Gulf states, but with some essential differences from Saudi Arabia. On the eastern fringe of the Arabian Peninsula, tribal societies emerged around twelve ruling families, each exercising authority over a small territory. These entities became the basis for differentiation from both the social and political forces within the Peninsula, and those on the opposite shores of the Gulf. The character of these political societies was shaped by a number of factors: the personalities within ruling tribal families; the intra and inter-dynastic struggles which they managed to contain; territorial disputes; and sectarian, tribal and even ethnic differences. The peculiarities of the economic resources at each tribe's disposal, and the way in which each managed to utilise these resources through economic and political intercourse with the interior and with maritime states, also contributed to the personality of the political forces in this region.[12] The patterns of trade defined the survival problem of these small states in terms of threats of absorption and domination by neighbouring or other powers.

Through the nineteenth and twentieth centuries, the shaykhs of the Gulf proved capable of playing off the Turks, against the Persians, the British, the Saudis and others.[13] In the end they fell under British protection which contributed greatly to their survival during the 1920s and 1930s as well as to the relative stability of their borders.[14] The British did not interfere with the internal ruling structures of the shaykhdoms but, in general, took responsibility for the conduct, though not the content, of their foreign relations. The effect of this policy was that the Gulf remained an independent sphere of decision-making and activity whose ruling structures remained intact in their traditional form even after the British withdrew from 'east of Suez'.

The British withdrawal left a 'vacuum' in the region that threatened the stability of inter-state relations in the Gulf. The small Gulf states have dealt with this problem largely on a bilateral basis with Saudi Arabia: getting it to aid in their security, while at the same time attempting to control the Saudi relationship by looking for support from outside powers. Little in the basic domestic political structures of the small Gulf states had changed, although each had grafted on aspects of the Western model. But the demands upon these political structures altered dramatically in the first decade following British withdrawal and have continued to do so since. The fragility of these internal structures led to the promotion of a federation of these states as both an immediate measure of reassurance, and as an attempt to ensure long-term political stability. This exercise, only partially successful, resulted in the formation of the United Arab Emirates.[15]

The strategy for survival of these small states, it has been thought, would be to institutionalise some sort of community of all the states in the region. Within such an arrangement states that are seeking both to prevent upsets to internal security, and to fend off external threats, would develop 'formal or informal institutions, sufficiently strong and widespread to assure peaceful change among members of a group with "reasonable" certainty over a "long" period of time'.[16] There have been several difficulties with this 'best' of all options, including disharmony among the states in the region as to objectives, sudden changes of government and policy, and disagreement as to the source of threats. These difficulties have meant that the idea of a Gulf community, which contains within it political, economic and military aspects, has not been a workable one – and was, indeed, closed off with the outbreak of the Iraq–Iran war in 1980. Instead the states with similar societies, historical backgrounds and political structures – i.e., Saudi Arabia and the small Gulf states, have haltingly moved towards the creation of an organisation which calls for integration in the social, economic, legal and administrative fields.[17] The Gulf Cooperation Council (GCC), ratified in March 1981, is merely one step along the road towards such a community.

To add to their internal difficulties, these states are also vulnerable to attack from their Gulf neighbours, have need of the assistance of outside powers, and, except for Oman, fear the internal and external political consequences of open reliance upon the US.[18]

It was clear from the reactions of Iran, Iraq and Saudi Arabia to the British announcement of withdrawal that each had ambitions concerning the Gulf. Both Iran (1968) and Iraq (1970) made approaches to the other littoral states to establish cooperation in military matters which included an agreement governing the Strait of Hormuz.[19] The objective was ostensibly to reduce the possibility of a non-Gulf power intervening in the area.

Both Iran and Iraq increased their military capabilities, so becoming the eminent and competing powers in the region. Both exhibited attitudes towards the Gulf that made Saudi Arabia and the other states suspicious that they might have hegemonic ambitions.[20] None of them, including Saudi Arabia, even in concert with the other Gulf states, could withstand pressures from Iraq or Iran except with the help of a major power.

The attitude of the GCC states towards Iraq was long-standing in that Baghdad had inherited Ottoman policies towards the Gulf. This is the background to the territorial disputes involving Iraq with Kuwait, Iran and Saudi Arabia. The treaties or agreements that resulted from the resolution of some of these disputes did not always put an end to the issues involved. Kuwait has received the brunt of Iraq's attentions, whereas Saudi Arabia has been fairly free of conflict with Iraq. This was due, in part, to the distraction of Iraq by internal problems, and to conflicts with Iran regarding claims of Iranian involvement in Kurdish discontent, Iranian interference with Iraqi ships on the Shatt, and the cooperation between Israel and Iran which complicated Iraq's security perceptions.[21]

Iraq as a secular, radical state, had long been a pariah among the states of the Gulf. But with the fall of the Shah from power in Iran and the outbreak of the Iraq–Iran war, the effect has been a realignment of states in the region which brought Saudi Arabia and Iraq close enough to each other to result in the signing of a security agreement.[22]

Before this seismic event in the Gulf, there had been in the second half of the 1970s numerous bilateral negotiations among Gulf states on internal security.[23] By 1978, urged on by the accidental assassination of the UAE Minister of State for Foreign Affairs, Said al-Ghobash in 1977, the majority were ready to cooperate in measures designed to combat terrorism and sabotage in the Gulf. The means agreed upon was the sharing of intelligence information and in general setting up the processes whereby these measures could be made effective. After the Shah's fall, the abrogation of CENTO (which was non-functioning in any case) and threats by the US to defend the oil-fields and international shipping in the Gulf, consultations were carried on among the Arab Gulf states, culminating in the establishment of the Gulf Cooperation Council (GCC) of May 1981.[24] It included all the Arab Gulf states except Iraq, though some have suggested that it is an unwritten member.

The GCC, however, is not a military defence treaty. Formed at the initiative of Saudi Arabia, it does create a loose political framework whereby the members can co-ordinate foreign and economic policies. However, military policies have proved most difficult to co-ordinate. Saudi Arabia has signed bilateral military treaties with its small neighbours but it is difficult to dispel the suspicion which the smaller states have regarding the larger states as to their security. If control over their own military is somehow diluted, their sense of vulnerability is heightened.

There appear to be a number of positive aspects to the GCC which may give a better than even chance of its survival and success. It is a homogeneous unit. There are considerable similarities in the social structures of the member-states – each state is ruled by a Sunni ruling family which is attempting to modernise its society; the indigenous populations are comparatively homogeneous (large populations of migrant workers are kept isolated, enjoy few political rights and are not allowed to become citizens); the economies are a mixture of free-wheeling private enterprise, laced generously with provisions for the welfare of its citizens; and there is a general awareness among the ruling families that social developments would generate discontent among the population. All the ruling families are concerned with the security of their regimes which is conceived in terms of stability, in general, and specifically with the legitimacy of their governments and ruling families. Policies and their implementation become the direct outcome of negotiations by the rulers themselves.

Looking briefly at one of the smaller, and perhaps, the most vulnerable of the Gulf states, the United Arab Emirates (UAE), we can see in microcosm the balancing act that is required for survival.[25] The UAE is sensitive to its weaknesses – that is – it consists of seven small shaykhdoms that have federated but have continued to pursue their individual interests within the confederation. Their ability in the past to pursue their interests which was due primarily to the British umbrella, now is due to their ability to balance the interests of one neighbour against another. Being located at the southern end of the Persian Gulf, the UAE member governments have traditionally seen the threats to themselves emanating from Saudi Arabia and Iran. Added to these longstanding concerns, new fears arose in the post-war era both from more radical elements of the Arab nationalist movement, and from radical activities of successive popular front groups based in the Dhofar province of Oman that were hostile to the traditional Gulf governments. The People's Democratic Republic of Yemen (PDRY) has been seen as a strong supporter of these groups. In one sense this threat has been reduced by the defeat of the rebellion of Dhofar province against the Omani government in the 1970s. But it has been supplanted by the new and different threats emerging from the Iranian Revolution, with its appeals across a wide range of social issues – corruption, questionable development projects, and conspicuous inequalities of wealth.[26] In these circumstances, the UAE and its member governments have had to pursue security in a way that will not encourage any of its neighbours to impinge upon its independence or integrity. Though the UAE is suspicious of Saudi designs on the region, its leaders recognise that they have little chance to remain independent without Saudi support. Even with Saudi support, a determined attack from either Iraq or

Iran could not be successfully resisted. In order to bolster its position, it accepts aid, as do the others in the GCC, from the US in the form of 'over the horizon' support.[27]

On the other hand, Kuwait's security has been more directly threatened by Iraq. At the centre of the GCC security problem is that it was formed as an attempt to mitigate against members' lack of effective defence against a spill-over from the Iran–Iraq war which might also contribute to superpower intervention. But a central difficulty is that members have different perceptions of the threats to themselves, and this presents problems for devising a common strategy. GCC members have spent heavily on a range of sophisticated weaponry and though not having a common military agreement, they have held joint military manoeuvres. Military cooperation among the GCC states has developed slowly. In October 1983, the Council announced its intention to create a rapid deployment force for the defence of the member-states in time of crisis. The assumption was that if the flow of oil were blocked within the Gulf the crisis would be handled by forces within the region without the involvement of the USA[28]

Within the GCC there are differences not only in defining the threat but also the means by which the threat is to be countered. Oman has accepted an undisguised role for the US in preserving its security. It has allowed the US basing and pre-positioning facilities, though reportedly only for responding to a Soviet threat. Saudi Arabia's population size in relation to some of its neighbours, its political and diplomatic prowess, its wealth, and its military capability, are not enough to deter hostile regional pressures. To strengthen its position, Saudi Arabia seeks a disguised relationship with the US and an 'over the horizon' US presence. Saudi Arabia has a shared security interest with the US in the Gulf, but so far it has not allowed either a pre-positioning arrangement with the US or basing facilities. Its military modernisation programme is heavily dependent on the US. The structure and training of its military is such as to facilitate co-ordination between Saudi Arabia and the US in time of crisis.[29] Bahrain lies off the Saudi coast and is tied to it by a causeway. Its size and limited capability for defence ensures that, like the UAE and Qatar, its foreign policy stance is compatible with Saudi Arabia's. Kuwait is the one member of the GCC that has diplomatic relations with the Soviet Union. It is extremely vulnerable to possible ground and air attacks from Iran and Iraq because of the concentration of its population and economic facilities in close proximity to Iran and

Iraq. The nature of its insecurity has induced Kuwait to seek better relations with both the US and the Soviet Union.[30]

When the GCC was established, what was conspicuous by its absence was any military component to the organisation. Though the stated aim of the GCC was to move towards integration, as is often the case, it is difficult for states to alienate control of decisions regarding their military. The Gulf states are certainly not unusual in this regard. Each state has developed its military forces independently. Each has shown an interest in diversifying its sources of supply and training – usually British, French and US. And since the oil price rises and the outbreak of war in the Gulf, these states have expanded their forces by use of foreign military advisers and troops. With the growing tension in the Gulf the inclination of the small states was to prefer bilateral defence relations with Saudi Arabia. This preference continues to underpin the small GCC members' approach. It is co-ordination, not integration, which they prefer. The problem for them is the disparity between Saudi Arabia and themselves and the likelihood of Saudi values and aims dominating any such integrated body. The idea of Saudi forces operating on their territory is distasteful to them, and is precisely what these small states have avoided historically.

Hence, in October 1983, the GCC held joint military manoeuvres and have continued to pursue this route.[31] The GCC Defence ministers have preferred to create a separate, smaller RDF which could protect the Gulf.[32] The idea of a force of 100 000 for regional defence against threats from the littoral states, proposed by Oman, has been acted upon and ratified in October 1983. Its purpose is to eliminate the necessity to call upon the US to intervene in the Gulf if the flow of oil should be blocked. As regards clear passage through the Strait of Hormuz, the GCC have declared that to be an international responsibility.[33]

It is clear that the security dynamics of the Gulf complex are defined by these issues: conflicting territorial claims, differing strategic objectives, and ideological conflicts. Economic and military capabilities within the region vary considerably. These differences in many cases tend to be resolved by strategies of alignment at the regional level. But the need for independent military capability, and the level of sophistication of the modern weapons required, tends to bring superpower involvement into the region.

THE PAKISTAN RELATIONSHIP

In the regions of the Middle East and South Asia, the security policies pursued by governments are largely defined by a matrix of three factors: domestic uncertainties that undermine the stability of civil society, interstate rivalries, and the impact of superpower strategies concerning the shape and content of the international order on the structure of regional relationships. In this context, what has been the content and character of Pakistan's relationship with the Gulf states, in particular, with Saudi Arabia?

Since 1947, when Pakistan came into being, it has had good relations with Saudi Arabia which welcomed the formation of an Islamic republic.[34] As a state based primarily on the commonality of Islamic religion, Pakistan sought closer relations with other Muslim countries in the Middle East. It nevertheless had divergent foreign policy needs. Though it identified with Arab opposition to Israel it avoided getting involved in inter-Arab disputes. Pakistan's involvement with the Baghdad Pact brought a cool response from many of the Arab states because the Pact was seen as an instrument of Western colonialism.

Pakistan's economic and security involvement with the Persian Gulf predates the secession of Bangladesh, but became more significant after the loss of East Pakistan, when its foreign policy orientation shifted towards the Middle East. India, however, still remained the primary threat, with Kashmir and the fear of dismemberment the main concerns. India had the basic elements of national power: population, economic resources, industrial capabilities and technological sophistication. Its defence capabilities were considerably augmented by its ties with the Soviet Union as a principal supplier of high technology arms. The Soviet Union, in addition, has assisted India in the development of industries for the production of major weapons systems.[35] Pakistan remained particularly vulnerable in its ability to sustain a long military campaign. The funding of such a campaign would be an onerous burden on its finances, and with the high rate of expenditure of material associated with modern warfare, the supply of munitions and equipment would be a problem. Pakistan has a further problem in that it has only one port, Karachi, which is vulnerable to air strikes from India, as are the major cities and communication routes of Pakistan. Pakistan's relations with Iran, under ideal circumstances, would be a route through which material could be supplied to Pakistan during war and whose breadth of territory would allow a defence in depth.[36]

In the circumstances of war, the substantial financial resources of Saudi Arabia could be a critical factor in funding a Pakistan war effort, and the availability of compatible, US-supplied war material in both countries would enhance Pakistan's ability to sustain such an effort. A suggestion of this factor can be seen in Saudi aid to Pakistan during the 1965 Indo–Pakistan war.[37] During the war, the US had imposed a ban on the shipment of arms to both Pakistan and India, with Pakistan being the more vulnerable to a cut-off of arms. Saudi Arabia made a modest contribution in response to Pakistan's request for arms, as did Turkey and Iran who supplied material, and served as conduits of arms purchased from Europe. The latter also gave financial assistance. However, this support did not in any way alter the military balance.[38]

There are other points of utility for Pakistan in the relationship with Saudi Arabia, such as the benefits of links with a country as important as Saudi Arabia in Western circles and particularly to the US. To have the political and diplomatic support of Saudi Arabia in international fora is an asset especially in relation to the US. Perhaps this is how the Fahd Declaration of December 1980, which underlined Saudi Arabia's full support of Pakistan if attacked, should be understood.[39] It is probably an example of Saudi diplomacy directed against Soviet pressure on Pakistan from Afghanistan.

The Pakistani relationship with Saudi Arabia began to deepen from the mid-1960s, when large numbers of Pakistani professionals and technicians were contracted by the Saudi government to assist it in its development programmes.[40] By the early 1970s, the practice was well established of professionally trained Pakistanis participating in Saudi economic programmes. The size and type of these programmes were such as to require increased numbers of foreign skilled and unskilled labour, of which Pakistan has been and continues to be an important supplier. The estimates given of Pakistani migrant workers employed in the Middle East can vary considerably. Some of the figures derive from estimates taken from government sources and licensed recruiting agents. It is believed that a considerable number of workers have found employment in the Middle East without the use of official channels or licensed labour agents.[41] A recent estimate of Pakistani migrant workers in Bahrain, Kuwait, Oman, Qatar, Saudi Arabia, and the UAE placed the current number of 2 829 000, a rise of 262 per cent over the 1975 estimate of 1 081 000.[42] In recent years, Saudi Arabia has attempted to attract migrant workers who, it believed, would be less likely to develop the desire for permanent residency – hence, migrant labour from South Korea, Philippines and Thailand.

The profile of the broader relationship between Pakistan and the Middle East points out the particular needs of Pakistan. It has a large population – around 92 450 000 – a weak economy and a chronic balance of trade deficit with oil-producing Arab states, particularly Saudi Arabia. Pakistan has been able to offset to a certain extent the effect of this on its balance of payments by the Pakistanis working in the Middle East sending home their remittances – amounting in 1982–3 alone to about $3.1 billion.[43]

Saudi Arabia, Kuwait and the UAE are important trading partners of Pakistan, ranking with Japan and the USA. Saudi Arabia ranks as Pakistan's second largest export market, with the UAE fifth. About 30 per cent of Pakistan's imports (nearly wholly petroleum and petroleum products) come from Saudi Arabia, Kuwait and the UAE.[44] Pakistan is largely dependent on oil from these sources.

The military linkages between Saudi Arabia and Pakistan are now considerable and serve to reinforce each other's security capability. In the latter half of the 1960s this relationship deepened as the result of a worsening regional situation in the Middle East marked by the growing military power of both Israel and Egypt and culminating in the 1967 Arab–Israeli war. In the south and southwest regions of the Arabian Peninsula the already unstable conditions deteriorated further in the precipitous withdrawal of British forces in 1967 from Aden, leaving in its wake a radicalised South Yemen government. In addition tribal discontent was simmering in Oman; there was the growing independence of the PLO with its attendant influence on the Palestinians residing in the Gulf area; the arms race between Iran and Iraq encouraged by the forthcoming withdrawal of Britain from the Gulf; and the unreliability of US arms sales to Saudi Arabia in the aftermath of the 1967 Arab–Israeli war. In response to these worsening conditions Saudi Arabia embarked upon a programme of modernising its military capabilities. It is during this period that Pakistan as well as the US, Britain and France became involved as consultants in the expansion and modernisation of Saudi armed forces. In August 1967, a defence pact was signed between Pakistan and Saudi Arabia in which Pakistan's participation as advisers in the military modernisation programme was assured.[45] Since then, because of a shared Islamic culture and complementary capabilities, military cooperation has become a significant aspect of their relationship. With the oil price rises in 1973–4 the funds available to Saudi Arabia were astronomical, allowing extravagant expenditures on the military.

The relationship between Saudi Arabia and Pakistan had already become closer before the military coup which brought Zia-ul Haq to power. When Zia introduced a much stronger Islamic element not only into the formulation of his domestic policy, but also into his foreign policy, ties with Saudi Arabia increased.[46] Pakistan was already giving training to the Saudis both in the use of equipment and logistics, assisting in the training of pilots both in American and French aircraft and training Saudi naval midshipmen in Pakistan. With the outbreak of the Iraq–Iran war in 1980, a real military threat to Saudi security developed from the Gulf, and this brought about a second round of high technology military purchases. This was far beyond Saudi capability to absorb, construct, service and man. Thus the Saudis were forced to make considerable use of foreign personnel from many countries in the military development programme and in the defence of their security. In 1976, Pakistani troops were part of the Saudi contingent of the Arab Deterrent Force sent to Lebanon by the Arab League. By 1979, around 5000 Pakistani troops were said to be operating in Saudi Arabia. In 1980, it was reported that about 3000 Pakistani technicians and instructors were in Saudi Arabia as advisers. In 1981, as many as two divisions were claimed to be in Saudi Arabia and it has been reported that Pakistani troops were deployed near the Saudi border with South Yemen. It is reported that Pakistanis perform security functions in the oil-fields of Al Hasa Province. Pakistani officers serve in both the Saudi army and National Guard. There are approximately 20 000 foreign contract military personnel in the country. Of these, Pakistani troops comprise a substantial number, varying in estimates from 7000 on up.[47] There are additional Pakistani military in comparatively substantial numbers in the UAE and Oman. Pakistan's military involvement in Saudi Arabia appears to be a substantial contribution in conjunction with the Saudi's 35 000 man army,[48] to the defence of the country.

Several observations need to be made to indicate more fully the possible nature of the relationship. In terms of training Saudi military, Pakistan has not had a clear field. The US has trained the lion's share of Saudi officers, and Pakistan has trained most of the remainder. Though the 1967 military agreement was renewed by Saudi Arabia in 1980, the following year, an agreement was signed with Turkey to train Saudi officers in the future.[49] This appears to be part of a more general shift among Gulf states to diversify their sources of foreign personnel for development needs.

Iran's territorial propinquity gives it an importance to Pakistan that is lacking in Pakistan's relations with Saudi Arabia. Pakistan not only shares a long, vulnerable border with Iran but an ethnic minority group – the Baluchis. Neither country has any territorial claims against the other but both are vulnerable to the possible disintegrative effects of conflicts within each other's territory. Evidence of the latter has been Iran's concern with the effects of the Indo–Pakistan wars and the Baluch insurgency of 1973–7 on the territorial integrity of Pakistan.[50] On each occasion, Iran supported Pakistan either by offering mediation, or supplying arms and aid or giving full military and diplomatic support or, during the insurgency, sending Iranian-piloted combat helicopters to raid Baluch camps.[51] The current Iraq–Iran war poses similar problems for Pakistan. If Iran were defeated, it could lead to separatist movements which could easily spill over into Pakistan itself. A victorious Iraq would then be able to dominate the Gulf.[52] The question would arise as to what advantage Pakistan would gain in becoming embroiled in interstate conflicts in the region, as opposed to simply confining itself to military aid in domestic matters involving the defence of the regime or the territorial integrity of the nation. To be involved in Gulf regional strife could complicate the internal stability of Pakistan and distract from Pakistan's main security concern – India – and raise the possibility of a defence of its frontiers on two fronts.

Both Pakistan and Iran, in the late 1970s, have undergone changes towards theocratically influenced governments. Pakistan was quick to recognise the new Islamic Republic that emerged from the Iranian revolution and relations between the two countries were amicable. Events were to complicate matters when the Soviet Union became militarily involved in Afghanistan, precipitating the subsequent shift of Pakistan towards the US. This brought Iran's condemnation of Pakistan as an oppressed Muslim nation with close ties to 'the great oppressor power', the USA.[53]

Although a coolness in the relations between the two countries has developed on the political level, on the economic level, it has been business as usual. Pakistan has expended considerable effort, with some success, to increase its export trade in both agricultural and industrial goods to the Middle East. By 1982 about 30 per cent of its export trade was with the Middle East.[54] The relations between Iran and Pakistan abjure a commitment to the objectives and ambitions of the others in the region. The interest that remains common to both is the insurance that the central government in each country will be maintained in the face of pressures from ethnic minorities. In this sense Iran maintains moderately fair relations with both Pakistan and India, while Pakistan maintains good relations with both Iran and her enemies in the Gulf. It is unlikely that Pakistan would conduct its relations with Saudi Arabia in a manner that would complicate or harm its relations with Iran. Pragmatic economic considerations and the reality of the situation override the ideological differences that do exist.

With perceptions and values encased in a similar philosophical world view, Pakistan and Saudi Arabia have sought greater security through linkages on the economic and military levels. Insecurity is defined by the inherent instability of the regimes, and by the existence of external threats. The linkages that have been formed between Pakistan and other states in the Middle East have enabled them to increase their survivability in an essentially unstable and disorderly regional environment. After the loss of East Pakistan, Pakistan's impetus to carve out a role for itself in West Asia has led it to attempt to balance Indian and Soviet power in the region by its own strategic relations with China and the countries of the Middle East.

Pakistan has attempted to shelter its domestic political affairs, which are structurally fragile, from unwanted influences. Its involvement with the Middle East affords its economic and financial advantages which help reduce the divisive stresses in its society. The countries of the Middle East have welcomed the willingness of Pakistan as an Islamic nation to contribute its military expertise and forces on behalf of their security in a way that would not jeopardise their independence. The mutually advantageous linkages that have formed between Pakistan and the Middle East do contribute substantially to each other's security needs in a potentially explosive regional context.

Regional security comprises the relationship among states within a defined complex. India defines its regional security primarily in terms of its relationship with Pakistan and the adjustments to this relationship in response to the involvement of the superpowers and China. But there is some blurring of boundaries between regions especially in the economic domain. India has important economic linkages in its dependency on Gulf oil. It also has considerable involvement in the economic development of the Gulf states, primarily in civil construction and economic and technical cooperation.

Commercial relations with the Gulf states have provided a growing market for the sale of Indian manufactures, technology and expertise. Indian migrant workers in the region were estimated to number 360 000. For the Gulf states, India appears to have been a satisfactory outlet for the recycling of petrodollars by means of loans and investment in the public as well as the private sector.

Though these linkages have historical antecedents they are not sufficiently strong and do not contain the geopolitical dimensions that would cause India to redefine its regional concerns through an extension into the Gulf region. The geopolitical effects of the Gulf region on India are peripheral to its primary concerns except where a crisis in the Middle East either brings about an alteration to the strategic position of Pakistan, or brings the military presence of a superpower into the region.

The explosion of a nuclear device by India in 1974 gave a considerable push to Pakistan's efforts to acquire a nuclear weapons' capability. Three years earlier Pakistan's already precarious security position was shattered by the events in East Pakistan in 1971. While the secession of the Bengali province considerably simplified territorial defence, it indicated Pakistan's susceptibility to the pressures of its ethnic minorities for greater autonomy or even secession.[55] Though border defence was less demanding, what remained still posed formidable problems of defence: the lack of natural barriers on the frontier with India, the heavy concentrations of populations and industry in urban centres, the vulnerability of transport and communications to interdiction by air, and the lack of strategic depth. The acquisition of nuclear military power would give Pakistan a defence posture that could serve either to deter a potential aggressor, or as a final solution to an impending military disaster. There is another possible reason for going through the considerable expense of acquiring a nuclear capability, that is, to 'exploit its political utility' in negotiations with other powers.[56].

A clue to the utility of a nuclear bomb produced by Pakistan is in the suspected funding of Pakistan's nuclear programme by both Saudi Arabia and the UAE and earlier by Iran and Libya.[57] The possession of nucler weapons would present Pakistan with the opportunity to offer the Middle East a 'nuclear umbrella' against threats from Israel or from powers outside the region. A general involvement of this nature would add to the already considerable involvement of Pakistan in the security of the region without antagonising individual states or threatening the existing regional balance.

CONCLUSIONS

The central concern of this chapter was whether the scope and character of relations between Pakistan and the Gulf states were of such significance as to cast doubt on the integrity of the 'boundary of indifference' between the South Asian and Gulf security complexes. The situation is, of course, not static, and therefore some sense of trends is as important as a definitive judgement about the relationships at present.

The evidence here enables us to conclude with considerable certainty that the relationship between Pakistan and the Gulf Arab states is not yet of sufficient weight to compromise the traditional separateness of the two regional security dynamics. Neither Saudi Arabia nor Pakistan play primary roles in each other's security domains, and neither of them has made the other's principal security perspectives an important part of its own. Pakistan has every reason to avoid taking sides in the major Gulf disputes. To do so would not only stretch its resources in an area where its weight could not produce a decisive outcome, but also endanger its strategic rear. Saudi Arabia has no quarrel with India, no desire to add to its list of enemies, and no wish to draw outside actors too closely into its local environment. The desire on both sides to avoid too close an entanglement is clearly indicated by Pakistan's non-involvement in the GCC.

That said, we are clearly looking at a security relationship of considerable secondary importance. Although this relationship is certainly facilitated by the common third-party links of both principals to the United States, it also contains a substantial component of real joint interest. The common Islamic commitment creates a definite sense of belonging to a distinct, if diffuse, political community. For Pakistan and Saudi Arabia this sense gives each of their governments a stake in the political legitimacy and survival of the other. Additional reinforcement of interest pehaps occurs between them by the fact that each is trying to defend its own status quo within its local complex, and that both are doing so against considerable internal and external pressure, and from a position of marked relative weakness. Neither threatens the other, and each can provide the other with important and complementary supports: money, oil, and possibly military equipment from the Gulf to Pakistan, and a large quantity and wide range of human resources from Pakistan to the Gulf. The limitation of Saudi–Pakistan relations is best shown in their common concern about the Soviet presence in Afghanistan. For Pakistan, this presents pressure on a core interest of immediate concern, whereas for Saudi Arabia, the pressure is more distant, and less intense. As a consequence, developments in the region could elicit different responses from the two countries. Although most of these could be accommodated in the existing relationship, some, such as a Pakistani accommodation with the Soviet Union, might threaten it.

In terms of trends the picture is much less clear. There is no obvious or compelling set of developments pushing to transform these secondary links into primary ones, and it is equally easy to speculate that they will expand, contract, or stay roughly in their current form. This confusion reflects the volatile character of both the states themselves and the pattern of relations among them.

The cases for continuation along current lines rests on the ability of the states concerned to hold both their governments and their regional positions without recourse to major conflict. Since their policies are aimed at precisely this outcome, success will tend to reproduce itself unless interrupted by larger internal or external forces. The arms-length support which Pakistan and the Gulf Arab states provide for each other, is thus both dependent on, and supportive of, the status quo. To reduce the possibility of a contraction in the relationship Pakistan must avoid becoming involved in Arab internecine conflicts that could antagonise other regimes and increase perceptions of it as a regional threat. With the possible exception of sustained Soviet pressure on Pakistan, where their common concern is clear, neither side has either the depth of interest or the surplus resources to risk itself in a major conflict of the other. In addition, and precisely because of the endemic instability of governments in the area, and the different national interests involved, neither side can count on the durability of the other's support. Under such circumstances, ordinary prudence requires that states avoid both making deep commitments to each other, and relying too deeply on such commitments as are made.

The only promising avenue for expanding security links between Pakistan and the Gulf is the one that moves towards nuclear cooperation. It is possible to imagine a situation in which Pakistan in some sense sells 'shares' in its nuclear programme to Arab states in return for their material and financial support. Pakistan's motive could arise from the financial strain of competing in a nuclear arms race with India. The Gulf states might be interested in such an

'Islamic bomb' for a variety of reasons: to counter developments in Israel, to play local balancing and deterrent games amongst themselves, or to increase their leverage on the great powers, both friend and foe. This scenario is, however, highly conditional. To say that it is possible, or even probable, is a long way from saying that it is likely. But on the present evidence, it looks like being the only line of indigenous development capable of merging the Gulf and South Asian regional security dynamics, and we will explore it in more detail in Chapter 9.

NOTES

1. M. Z. Ispanani, 'Alone Together: Regional Security Arrangements in Southern Africa and the Arabian Gulf', *International Security*, vol. 8, no. 4 (Spring 1984) p. 154.
2. E. Hermassi, *The Third World Reassessed* (London, 1980) pp. 93–119.
3. G. M. Badr, 'A Historical View of Islamic International Law', *Revue Egyptienne de Droit Internationale*, vol. 38 (1982) pp. 5–6.
4. Ibid.
5. The following are useful sources on the history of Saudi Arabia: Christine Helms, *The Cohesion of Saudi Arabia* (London, 1981); David Holden and Richard Johns, *The House of Saud* (London, 1982); Robert Lacey, *The Kingdom* (London, 1982).
6. B. Tibi, *Arab Nationalism*, p. 63; David Long, *The Governments and Politics of the Modern Middle East*, p. 89 f; Helms, *The Cohesion of Saudi Arabia*, pp. 76–121.
7. Bahgat Korany and Ali E. Hillal Dessouki, *The Foreign Policies of Arab States* (London, 1984) p. 244.
8. S. Reza, S. Islami and Rostam Mehraban Kavoussi, *The Political Economy of Saudi Arabia* (London, 1984) pp. 31–2.
9. Holden and Johns, *The House of Saud*, pp. 198–286, 326–526; Lacy, *The Kingdom*, pp. 225ff; Anthony H. Cordesman, *The Gulf and the Search for Strategic Stability: Saudi Arabia, The Military Balance in the Gulf, and Trends in the Arab–Israeli Military Balance* (London, 1984) pp. 101, 148.
10. *Middle East and North Africa, 1882–3* (London, 1982) pp. 677–705.
11. Ibid., p. 245, Cordesman, *The Gulf and the Search for Strategic Stability*, p. 71.
12. The intensity of the trade within the Gulf is seen by the existence of Sunni communities on the Gulf shores of Iran and Shia communities on the Arabian littoral. See also John Duke Anthony, 'The Persian Gulf in Regional and International Politics: The Arab Side of the Gulf', in H. Amirsedeghi, *The Security of the Persian Gulf* (London, 1981) pp. 170–74.
13. Joseph J. Malone, 'The Historical Perspective', paper given at a Sympo-

sium on Iraq: The Contemporary Arab State, Centre for Arab Gulf Studies, University of Exeter, 2–6 July 1981.
14. Helms, *The Cohesion of Saudi Arabia*, pp. 284–9.
15. Holden and Johns, *The House of Saud*, pp. 300–302.
16. Richard W. Van Wagener, *Research in the International Organization Field: Some Notes on Possible Focus* (Princeton, 1952) pp. 10–11, quoted in H. A. Al-Ebraheem, *Kuwait and the Gulf* (London, 1984) p. 53.
17. Al-Ebraheem, *Kuwait and the Gulf*, p. 70.
18. Ibid., pp. 82–4.
19. S. Chubin, *Security in the Persian Gulf*, IISS (London, 1981) p. 154.
20. Securing its sea lanes out of the Gulf, in effect, the dominance of the Gulf, was a concern of Iraq. Maintaining claims to Bahrain, the Tumbs and Abu Musa would have had a similar result for Iran.
21. Anthony, 'The Persian Gulf in Regional and International Politics', p. 182.
22. Chubin, *Security in the Persian Gulf*, p. 154.
23. Ibid., p. 153.
24. Ispanani, 'Alone Together', p. 155; Al-Ebraheem, *Kuwait and the Gulf*, p. 68 ff.
25. Anthony, 'The Persian Gulf in Regional and International Politics', p. 174.
26. Ibid., pp. 193–4; Cordesman, *The Gulf and the Search for Strategic Stability*, pp. 432–8.
27. 'US Security Interests in the Persian Gulf', US House of Representatives Report, pp. 382–23, CIS–18, 16 March 1981; 'War in the Gulf', Committee on Foreign Relations, US Senate, Staff report, August, 1984, pp. 31–2.
28. *Asian Recorder*, Report No. 17517, 1983.
29. 'US Security Interests in the Persian Gulf', US House of Representatives Report.
30. 'War in the Gulf', Staff Report, Committee on Foreign Relations, US Senate, August, 1984.
31. *Asian Recorder*, Report No. 17517, 1983.
32. *Asian Recorder*, Report No. 16971, 1982.
33. *Asian Recorder*, Report Nos 17516 and 17517, 1983.
34. Hermann Eiltz, 'Saudi Arabian Foreign Poilicy toward the Gulf States and Southwest Asia', in Hafeez Malik (ed.), *International Security in Southwest Asia* (Eastbourne, 1984) p. 100.
35. 'Soviet Role in Asia', Statement of R. E. Kanet, House of Representatives Report, H381-1, 1983.
36. L. Ziring, 'Dissonance and Harmony in Indo–Pakistan Relations', pp. 1–18 and K. Prasad, 'Pakistan–Iran Relations', p. 144, *Punjab Journal of Politics*, vol. 6, no. 2, July–Dec. 1982.
37. P. B. Sinha, 'Pakistan–Saudi Cooperation', *Strategic Analysis*, vol. 4, no. 8, November 1980.
38. Prasad, 'Pakistan–Iran Relations', p. 144f.
39. As quoted in the *New York Times*, 11 December 1980, 'Any interference in the internal affairs of Pakistan would be considered interference or injury to the Kingdom of Saudi Arabia'. Quoted in Shirin Tahir-Kheli

and W. O. Staudenmaier, 'The Saudi–Pakistan Military Relationship: Implications for US Policy', *Orbis,* Spring, 1982, p. 159.
40. Eilts, 'Saudi Arabian Foreign Policy towards the Gulf States and South Asia', p. 102.
41. *Asian Recorder,* Report No. 16889, 1982.
42. I. Sergeldin *et al., Manpower and International Labor Migration in the Middle East and North Africa,* 1983.
43. *Annual Report,* State Bank of Pakistan, 1982, p. 142.
44. *Pakistan,* Grindlays Bank Group, February 1984.
45. Sinha, 'Pakistan–Saudi Cooperation', p. 357.
46. G. Dhanani, 'Pakistan and Saudi Arabia: An Alliance for Survival', *Punjab Journal of Politics,* vol. 6, no. 2, July–December, 1982.
47. *The Military Balance,* 1984–85, IISS, p. 107.
48. Cordesman, *The Gulf and the Search for Strategic Stability,* p. 361.
49. 'Pakistan, A Storm Centre on the Indian Ocean', Round Table Discussion, Friedrich Ebert Stiftung Research Institute, Bonn, 14 July 1981.
50. Prasad, 'Pakistan–Iran Relations', p. 144f.
51. S. S. Harison, 'Baluch Nationalism and Superpower Rivalry', *International Security* (Winter, 1980–81) p. 152.
52. P. B. Sinha, 'Gulf War and Pakistan', *Strategic Analysis,* vol. 4, no. 7, October 1980.
53. Statement by Richard Cotton, 'Soviet Role in South Asia', Subcommittee on Europe and Middle East and West Asian and Pacific Affairs of the Committee on Foreign Affairs, House of Representatives, 98th Congress; 1983, p. 485 ff.
54. Malik, *International Security in Southwest Asia,* p. 166.
55. R. W. Jones, 'Nuclear Proliferation: Islam, the Bomb and South Asia', *The Washington Papers,* vol. 9 (1981) p. 36.
56. K. Subrahmanyam, 'Pakistan's Nuclear Capacity and India's Response', *Link,* 2 December 1984.
57. G. H. Quester, *Nuclear Proliferation,* London 1981, p. 175, though Cordesman denies that there is any evidence of this, *The Gulf and the Search for Strategic Stability,* p. 768.

7 The Sino-Soviet Complex and South Asia

ROSEMARY FOOT

For the past twenty years or so, the Sino-Pakistani and Soviet–Indian relationships have been noted for their stability. Sustained by the hostility between Islamabad and New Delhi, between Beijing and Moscow, they have survived a number of regime changes in each of the countries, the break-up of Pakistan into two separate states, and international systemic changes in which multipolarity has replaced the bipolarity of the early cold war years. For reasons relating to the local dynamics and to the dynamics of the rift between the Soviet Union and China, the period under study will broadly be divided into three. In the first phase (1950–58) US penetration of the region stimulated a Soviet response and Soviet–American rivalry. The second phase (1959–72) witnessed Sino-Soviet rivalry as the most salient external force in the region. In the years following the emergence of Bangladesh to the early 1980s, however, the pattern of contention and the relationships it engendered have been disturbed, the full implications of which are still difficult to assess. Improvements in Sino–American and Sino–Indian ties, India's nuclear explosion and the Soviet intervention in Afghanistan are some of the factors that have contributed to the greater fluidity and uncertainty in relations in this final period.

AMERICAN PENETRATION OF SOUTH ASIA AND THE SINO–SOVIET RESPONSE

The United States was the first major power to enter the sub-continent after partition. The Soviet Union's atomic explosion in August 1949 and the establishment of a Communist government in

China stimulated a reappraisal of US strategic policy including that towards South Asia. Though Washington concluded that American resources should primarily be devoted to Europe and the Far East reflecting America's strategic priorities, the National Security Council agreed in December 1949 that since the 'non-Communist governments of South Asia already constitute a bulwark against Communist expansion in Asia, the United States should exploit every opportunity to increase the present Western orientation of the area'.[1] And, as a reflection of these needs, both Prime Minister Nehru and Prime Minister Liaquat Ali Khan were invited to the United States in 1949.

In a bid to acquire American arms, Liaquat, unlike Nehru, spoke out against Communism during his visit, stressed his country's strategic location and argued that its Islamic way of life meant that 'Communism was not likely to find fertile ground there'.[2] But Pakistan's major security concern was India not the USSR, and its preoccupation was to find the best means for resolving the many outstanding disputes it had with its larger neighbour, particularly that over the accession of Jammu and Kashmir. Pakistan's and America's interests nevertheless coincided even if the basis of these interests differed, and the relationship soon bore fruit. By 1954 Pakistan formally requested US military aid and training under the Military Defense Assistance Agreement and began to acquire the sophisticated arms thought necessary to secure its defences against India. In September Pakistan became a founding member of SEATO and one year later joined the Baghdad Pact, gaining the dubious distinction of being America's 'most allied ally' in Asia. In a number of areas, however, the relationship did seem to provide much of what Pakistani leaders wanted: in addition to receiving arms aid, Pakistan also received more than $730 million in US economic assistance between 1953 and 1958,[3] and US support in the United Nations for its stand on Kashmir.

Furthermore, and of major benefit to Pakistan, Indo-American relations were cool and unproductive for most of this period. Unresponsive to American calls to denounce Communism, and firmly wedded to a policy of non-alignment, India's policy was distrusted and misunderstood in Washington. America failed to take advantage of India's distress over China's military action in Tibet in 1950, or to capitalise on the enforced cooperation that emerged as a result of India's role at the Korean War truce negotiations. In the US Secretary of State's view, the spiritual faith and martial spirit of the Pakistani people made them a 'dependable bulwark against communism'.[4] India's non-alignment, peaceful diplomacy and secularism implied that they were not as reliable.

The American role in South Asia in the 1950s stimulated Soviet and Chinese interest in the area. Despite Stalin's two-camp view of the world and consequent suspicion of India's non-alignment policy, the Soviet press did draw distinctions between India and Pakistan as early as 1950, noting that the latter 'was to be the backbone and mainstay of American imperialist policy in the Middle East and South East Asia, in return for ... American arms and equipment'.[5] In 1953 there were reports of Anglo-American attempts to include Pakistan in the 'aggressive Middle East Bloc' and quotations from Nehru showing he deplored such moves.[6]

The first high level indication of a change in Soviet perceptions of India came in August 1953 when Malenkov, Chairman of the Council of Ministers to the Supreme Soviet, noted that India had made a serious contribution to efforts designed to end the Korean conflict, that Indo-Soviet relations were growing stronger, and that he hoped these ties would 'continue to develop and grow, with friendly cooperation as the keynote'.[7]

The impetus for the Soviet Union to develop its relationship with India was strong. In addition to its non-aligned policy, India was large, strategically located, politically stable and potentially powerful. Tangible Soviet support was not long in coming, first in the economic field and over the thorny issue of Kashmir. During the Khrushchev–Bulganin visit to India in November 1955, the two Soviet leaders offered aid and declared the Kashmir issue to have been settled by the people of the area themselves.[8] In 1957 Moscow vetoed a Security Council resolution proposing a plebiscite be held in the province. Trade between the two states also began to mushroom rapidly, rising from $1.6 million in 1953 to $94.6 million in 1958,[9] and ministers exchanged visits on a frequent basis. Between 1950 and 1960, for example, there were forty political and economic exchanges between Indian and Soviet officials, compared with eighteen with Afghanistan and only ten with Pakistan.[10] Soviet policy towards Pakistan was ambivalent, reflecting Pakistan's decision to join the Western military containing schemes. At times, Moscow would congratulate Islamabad, for example, for attending the 1955 Bandung conference, would offer economic assistance and to share technical knowledge in the peaceful uses of atomic energy,[11] but would accuse Pakistan of engaging in activity which threatened the security of the Soviet state. After the U2 incident in May 1960 the

USSR warned Pakistan that it risked Soviet nuclear retaliation if flights of this kind were again authorised to take place from Pakistani territory.[12]

In the 1950s, then, Soviet–American rivalry in South Asia was obviously exacerbating the divisions between India and Pakistan. International attention had become focused on Kashmir rendering compromise on the issue even more difficult to reach; American arms aid to Pakistan was upsetting the balance of power on the subcontinent; and Moscow's preference for New Delhi and threats of retaliation against Pakistan had raised the level of tension in the area.

China contributed less than the superpowers in the 1950s to divisions between India and Pakistan, wittingly or unwittingly increasing its options when at the end of the decade its relations with both Moscow and New Delhi deteriorated. Its rather distinctive approach towards South Asia probably derived from the competitive nature of Moscow's and Beijing's role in the newly decolonised areas, and from the presence of negative as well as positive elements in the Sino–Indian relationship. The negative aspects related to Tibet and to border delineation questions, and their presence may have contributed to China's relatively friendly approach towards Pakistan. Though Pakistan's decision to join Western military blocs dismayed Beijing, China issued few harsh propaganda statements, preferring to concentrate on the threat membership posed to Pakistan's independence and to the well-being of other Asian neighbours.[13] Between 1954 and 1956, China's premier, Zhou Enlai, made a number of favourable references to Pakistan at the latter's National Day receptions in Beijing,[14] and in 1955 at the Bandung conference of Asian and African nations, Zhou participated in a crucial discussion with Mohammed Ali, in which the latter assured the Chinese leader that Pakistan's membership of SEATO was not directed at China. Neither would Pakistan fight alongside the United States in any war against the People's Republic. Zhou recorded his grateful thanks for this explanation, which had increased the two states' 'mutual understanding', he said.[15]

In 1956, Prime Minister Suhrawardy and Zhou Enlai exchanged visits and recorded that there were no conflicts of interest between the two states. Even on the delicate issue of Kashmir, the Chinese government managed to avoid taking too controversial a stand, instead urging peaceful negotiations without outside interference. Indeed, Mao Zedong is supposed to have remarked to a Pakistani parliamentary delegation that it would have been better if other Communist countries had remained neutral also.[16]

But despite these attempts to control bilateral problems, good relations with China could not contribute significantly to Pakistan's most pressing security needs. Indeed, they ran the risk of undermining US–Pakistani relations if Washington chose to punish Islamabad for too close an association with Beijing. In addition, there was little two-way trade, China was not a source of development assistance in those years, and since it was not on the UN Security Council, could not play a role in the Kashmir dispute. The impetus for expanding ties was thus rather slight, in contrast to Sino–Indian relations which offered great opportunities for advancement.

China first came to appreciate the role of India in world politics during the Korean War when Indian officials were used (with mixed results) as a conduit for transmitting messages to the Western powers. In 1954, their bilateral relations were advanced when they resolved the difficult issue of the status of Tibet in China's favour, and jointly formulated the five principles of peaceful coexistence. India also steadfastly supported the PRC's assumption of the China seat in the UN Security Council.

India was a much more prestigious state than Pakistan and good bilateral relations could leave China relatively free to concentrate on its security concerns in East and South-east Asia where a hostile American presence loomed larger than in South Asia. Beijing could not make the kinds of economic offers that Moscow could bring to New Delhi. Neither was it willing to get drawn into the Kashmir issue; but visits exchanged between Nehru and Zhou were of great symbolic significance. Their friendly relations, based on the five principles, or *Panch Sheel*, were a model for all nations. They seemed to represent the triumph of Asian nationalism, and to herald a new political consciousness and maturity in a part of the world that for so long had endured an alien colonial presence. This spirit made both parties reluctant to face the more mundane problems of rival border claims.

THE EMERGENCE OF THE SINO–SOVIET RIFT

Towards the end of the decade, Soviet and Chinese foreign policy interests came into conflict and their world views began to diverge dramatically. Where Beijing developed a more militantly anti-imperialist policy in 1957 and 1958, drawing sharper distinctions between progressive and reactionary states, pressing governments to take a firmer stand against the Imperialist West, Moscow remained wedded to a policy of peaceful coexistence, and to a search for *détente* with the United States. Where China viewed Soviet technological achievements in space and nuclear weaponary as permitting the socialist bloc (and particularly its leader) to take a more confrontational stance with the United States, Moscow treated these developments cautiously, seeing them as opening a new era in international relations in which a strategic understanding with Washington might prove possible.

Neither the failure of the Soviet–American summit in May 1960, the Berlin crisis of 1961, nor the Cuban missile crisis, could halt this progress towards *détente*, which seemed sealed in the Partial Test Ban Treaty of 1963. China had argued that increased Soviet leverage should be put at the service of the socialist bloc, and had seemed to test Soviet willingness to use its power on behalf of its major ally during crises in 1958 over the Chinese offshore islands, and over the Sino–Indian border in 1959 and 1962. In each case China had found the Soviets to be wanting. In its view, therefore, it seemed plain that the 'revisionist' Soviet Union preferred 'collusion' with America in order to secure *détente* and to contain China and the revolutionary aspirations of Asia, Africa and Latin America. Moscow could no longer play a progressive role in any united front against imperialism.

By 1968, and as a result of the Soviet intervention in Czechoslovakia, China viewed Russia not simply as 'revisionist' but now as 'social imperialist'. Clashes on the Sino–Soviet border between March and September 1969 reinforced the view that Moscow ranked alongside Washington as a leading oppressor state and as China's major enemy. Indeed, in the aftermath of the Tet offensive in South Vietnam, the decision to 'Vietnamise' the war in South-east Asia, and evidence of a diminishing US presence in Asia, China perceived the United States as being on the defensive in world politics; as 'going downhill' and becoming 'less and less effective'.[17] Beijing began to refer to 'struggle' as well as 'collusion' between Moscow and Washington, where previously it had only spoken of their collaboration. Third parties, obviously, could exploit these contradictions between the two superpowers; and by 1971, it became apparent that China was willing to make a temporary accommodation with the United States against its more dangerous and vigorous enemy, the Soviet Union.[18]

In Moscow's view, China's behaviour over this period was both unrealistic and reckless. China did not understand the destructive power of nuclear weapons, nor the realities of the modern age. Mao, the supreme 'dogmatist', headed a military–bureaucratic dictatorship bent on expansion in Asia and determined to incite war between the Soviet Union and the United States. Containment of China thus became a major priority for the USSR, particularly since the Ninth National Congress in Beijing in 1969 confirmed the end of China's period of neglect in diplomatic activity. The border war (during which Moscow considered launching a pre-emptive strike against Chinese nuclear installations[19]) also suggested that the USSR should now rank the PRC as one of its major foes. In June of that year Brezhnev offered an institutional solution to the containment of China: the establishment of a collective security system in Asia, linked to an earlier plan for regional economic cooperation to involve the South Asian states, plus Iran and Afghanistan.[20] Coinciding with this, Soviet leaders also proposed that India sign with it a treaty of friendship and cooperation.[21] A common front against the 'adventurist' Chinese had to be forged, particularly in South Asia.

Sino–Soviet Rivalry and South Asia

Inevitably, the rivalry between the two Communist powers spilled over into neighbouring areas, and even beyond them. It was particularly acute in South Asia, mainly for geostrategic reasons and because of the re-emergence of unresolved issues between India and China which erupted in a border war in 1962. This conflict led to the formation of clear alignments in the region, with the Soviet Union and India forging close ties, countered by a weaker but still significant informal alliance between China and Pakistan.

The US role in South Asia during this period facilitated the development of these alignments. Washington's desire to enlist India in its anti-Chinese strategy severely weakened its relations with Pakistan, as did its growing preoccupation with the Vietnam War and technological developments that rendered forward bases in Pakistan of less importance to US military strategy. In many respects Washington's and Moscow's aims in South Asia ran parallel, as each tried to diminish Indo-Pakistani differences and to encourage the two contending states to concentrate on the threat from China. When the US Congress cut off all arms aid to the sub-continent in 1965 in response to the Indo–Pakistani war, the United States seemed relatively sanguine about the Soviet assumption of the role as major arms supplier to India and as peacemaker in the region, provided these policies contributed to the goal of containing China.

With relatively little effort, Moscow had by 1965 made substantial progress in achieving one of its primary goals in South Asia: to limit the American presence in the area. It also hoped to demonstrate to the developing world through the consistency of its efforts in New Delhi that it was a friend to rely on; and, through the closeness of Indian and Soviet positions on international issues, to show that it was a natural ally of the Third World, where China was not.

The other major Soviet goal of minimising Chinese influence in Pakistan was to prove more problematic, because Pakistan's and China's coincidence of interest grew as a direct result of the Sino–Indian conflict. Islamabad now had a partner who seemed willing to give material and political support in the event of war with India, as Foreign Minister Bhutto had implied in July 1963 when he stated that an Indian attack on Pakistan 'would also involve the security and territorial integrity of the largest state in Asia'.[22] For Beijing, on the other hand, Pakistan could provide the breach in an arc of hostile powers surrounding China; an ally that could prevent the consolidation of Soviet power in an area of vulnerability.

In many respects geographical issues are the key to an understanding of the Communist powers' relationships in South Asia and help to explain both the durability of Soviet and Chinese interests in the area, and how penetration of the region was accomplished. South Asia is adjacent to areas where Moscow and Beijing are at their weakest politically. Minority peoples are located there, they are distant from central government control, and they lack clearly delineated boundaries in some instances. During the Second World War, the success of German attempts at subversion in Soviet Central Asia left an indelible impression on the government in Moscow. Similarly, Chinese leaders have often alluded to the recurrent problem of 'local nationalism' and subversive activity in Xinjiang and Tibet.[23] Evidence of Soviet influence in Xinjiang and subversion on both sides of this border are well-documented. Indian influence in Tibet, where insurgency and unrest have been a spasmodic problem since the Chinese takeover in 1950, has always been a source of tension in the Sino–Indian relationship. But the vulnerability of border regions is not just a problem affecting the major powers; ethnic and territorial disputes affect all states of the area, contributing to the sense of insecurity and providing opportunities for the major powers to apply pressure. As noted above, Moscow gave all states of the area, contributing to the sense of insecurity and providing opportunities for the major powers to apply pressure. As noted above, Moscow gave support to India over the unresolved question of Kashmir in the 1950s, and has also given support to the Pathans government's calls for independence or autonomy for the Pushtus in Pakistan. There have been rumours of direct Soviet support to the independence movement in Baluchistan and of similar Chinese support to the Mizos and Nagas in north-eastern India.

In 1959 tensions in the Himalayan region finally came to a head. That year saw the first clashes on the Sino–Indian border and a rebellion in Tibet, as a consequence of which the Dalai Lama and thousands of his followers fled to India. The border clash, evidence of collusion between the CIA, Chinese Nationalist agents and Tibetan insurgents who were conducting subversive operations from Indian soil, coincided with a further deterioration in Sino–Soviet relations, and the collapse of the Chinese economy as a result of the failure of Great Leap policies. These combined events make the period one which is vital for an understanding of China's preceptions of South Asia.[24]

The Soviet Union's neutral stance on the border clash and its urging of Beijing and New Delhi to engage in talks fuelled the Sino–Soviet rift. Policies towards India in fact became an important symbol of their dispute, as Moscow accused Beijing of initiating the fighting in 1959, and chose to withdraw its aid from China when small quantities of Soviet arms and economic assistance began to arrive in India.[25]

These shifting policy alignments received a major boost in 1962 which witnessed serious unrest on the Sino–Soviet border[26] and fighting between Chinese and Indian troops in the Himalayas. Given the growing antagonism in Sino–Soviet relations, and despite the complications introduced that year by the Cuban missile crisis, Moscow finally resolved on a neutral position towards the Sino–Indian war. Moreover, though Britain and the United States were India's major weapons suppliers during the conflict, Moscow also provided arms. Since 1961 it had been shipping transport planes and helicopters — including crews for training in mountain warfare — and it continued this into 1962. In November 1962 at the height of the conflict Moscow honoured an earlier agreement to supply MiG 21s and to establish a plant to manufacture frames and engines in India.[27] As Neville Maxwell has written, India moved from non-alignment to a kind of bi-alignment with the Soviet Union and the United States against China. When Nehru informed Soviet leaders of his request to Washington for military help, he was reportedly told that Moscow 'understood both the request and the need for it'.[28]

As a consequence of Sino–Indian and Sino–Soviet hostility, the enhancement of ties with Pakistan became imperative for China. Pakistan, having become totally disillusioned with its Western alliance, believed the time had also come for it to reappraise its foreign policy stance. The mutuality of Chinese and Pakistani interests ensured a rapid improvement in relations. In 1963 the two signed border, trade and air agreements; in 1964 the Chinese offered their first interest-free loan and began high level military contacts. Chinese military assistance began in 1966 and, because of the 1965 US embargo, Beijing soon became Islamabad's principal arms supplier.

The major change in China's position on Kashmir occurred during Zhou Enlai's visit to Pakistan in February 1964. In the joint communique issued after the talks, China and Pakistan 'expressed the hope that the Kashmir dispute would be resolved in accordance with the wishes of the people of Kashmir as pledged to them by India and Pakistan',[29] a stance which China maintained until the late 1970s. In exchange, Pakistan defended China's position in the Sino–Indian border negotiations and supported the idea of holding a second Bandung conference – a primary political goal of China's since such a conference would challenge India's dominance of the Non-aligned Movement, and, as in 1955, probably would exclude Moscow from its deliberations.

In later years when Moscow discussed regional economic and security schemes with Pakistani leaders, Pakistan's political support for China proved even more valuable. As China noted,

> when a superpower, flaunting the banner of 'regional economic cooperation,' peddled a pernicious 'system of collective security in Asia' in a vain attempt to control the Asian countries, further push its aggression and expansion in Asia and oppose China, the Pakistan government and people firmly rejected the proposal.[30]

Perhaps China's most significant political moves in support of Pakistan occurred during the 1965 and 1971 Indo-Pakistani wars. During the first conflict, which spread from the Rann of Kutch in April to Kashmir in September, Beijing supported 'Pakistan's counterattack in self-defence against India's armed provocations.'[31] China next linked the Sino–Indian and Indo–Pakistani conflicts and issued a protest note to New Delhi demanding that it 'dismantle all aggressive military structures on the China–Sikkim border, withdraw its aggressive armed forces and stop all its acts of aggression and provocation against China, in the Western, Middle and Eastern sections of the Sino–Indian border. Otherwise India must bear responsibility for all the consequences arising therefrom.'[32] Just over a week later, China issued another ultimatum, giving New Delhi three days in which to dismantle its military structures. Two days later, the Chinese set back the time limit to September 22, the day that Foreign Minister Bhutto informed the Security Council of Pakistan's acceptance of the ceasefire resolution.

Assessments differ as to the value of China's intervention in the conflict, some analysts seeing it as vital to Pakistan, others doubting whether Beijing would have intervened militarily and finding no conclusive evidence that China's actions significantly improved Islamabad's position in the subsequent negotiations.[33] China did, however, have a considerable impact on India's strategic thinking and on New Delhi's perception of the threat Beijing posed. This prompted Indian leaders to sign the Indo-Soviet Treaty of Friendship in 1971 in order to reduce the possibility of China's intervention in future sub-continental wars. Furthermore, China's ultimata to India were a highly visible form of political support to Pakistan, and served to solidify the foundations of their alliance.

China did not adopt a similar posture in the 1971 war. Nevertheless, Beijing did continue to give Pakistan public support at a time when international public opinion was highly critical of West Pakistani actions in the East, and despite the damage caused to China's image as a supporter of national liberation movements. China was much more ambivalent in 1971, believing that Yahya Khan should negotiate an end to the struggle rather than seek a military solution, and privately warning Pakistan that the flight of refugees into India would spread the conflict and give India 'a reason or excuse', for open involvement.[34] China gave no indication, however, that it would intervene if India attacked Pakistan. Chinese leaders did give support to Pakistan in its 'just struggle to defend national independence and state sovereignty' but there was no expression of support for its 'territorial integrity.' On 16 December, the day that General Niazi surrendered in East Pakistan, China protested to India over the violation of its territory, but issued no ultimatum.[35] Chinese diversionary activity would obviously not be forthcoming and anyway, was now too late.

Part of the reason for the greater Chinese circumspection in this conflict relates to the terms of the Indo–Soviet treaty and explicit threats from Moscow. *The Anderson Papers* have reported that the

Russian military attaché contacted his Chinese opposite number in Nepal and warned of a Soviet reaction if China intervened in the conflict.[36] Moscow may also have moved ground and air forces into position along the Xinjiang border and trained missiles on Chinese targets. In addition, the Soviet ambassador in New Delhi reportedly told Indian officials not to be concerned about a Chinese attack since this would result in Soviet diversionary action in Xianjiang.[37] If India had reason to worry about a two-front war in 1965, it seemed that China did too from 1971.

Yet even without these threats, it seems doubtful from the evidence that China would have played a military role in the 1971 conflict. China pressed the Yahya government to exercise restraint in East Pakistan and pointed to the dangers inherent in his confrontational policy. The Bengali movement's links with New Delhi precluded Beijing's support for their struggle, but Yahya's policy of repression also ruled out China's active support on his side, not only because his actions encouraged major power intervention in the war, but also because, as Zhou Enlai said in 1972: 'Our assistance is for repelling aggression and not for suppressing the people.' That is why Chinese assistance 'remained limited in the past and why China could not do more'.[38] China's ties with Pakistan were solid but not unconditional.

Moscow's response to its deteriorating relationship with Beijing and to the developments in the region was more complex than China's, where hostilities and friendships were clear cut and seemingly irreversible. Though New Delhi was of paramount importance to Moscow as a counterweight to Beijing, the decline of US interest in South Asia opened up opportunities for improving ties with Pakistan. As noted previously, such a move could achieve two of Moscow's primary objectives: the reduction or even elimination of Western influence in Pakistan, and the minimisation and containment of China's influence in the area.

This dual strategy, involving friendly relations with both India and Pakistan, was to prove difficult to manage. Moscow probably believed it could adopt this approach successfully because of the solid economic and military ties it had forged with India: between 1954 and 1975 India received, for example, some $1263 million in grants and credits from Moscow, 18 per cent of all Soviet assistance provided to non-Communist developing countries over this period. Arms transfer were also substantial, reaching $1375 million in value between 1965 and 1974 compared with $41 million from the United States, $78 million from Britain and $39 million from France.[39] India was also the only country outside of the Warsaw Pact area to have licensed manufacture of the MiG. Furthermore, by 1965, India had become Moscow's largest non-Communist trading partner, the USSR taking an increasingly large share of New Delhi's manufactured and semi-manufactured goods.[40] Given this solid basis to the relationship, Moscow probably calculated that it could branch out in the region without damaging relations with India. Pakistan's shift from alliance with Washington to an alliance with Beijing could then be halted.

Three issues restricted the improvement of relations between Moscow and Islamabad: the Soviet attitude on Kashmir; Soviet arms transfers to India; and Pakistan's membership of SEATO and CENTO. Kashmir was the first area to show signs of change as Moscow started to advise New Delhi in 1962 and 1963 to adopt a more flexible attitude during the negotiations then in progress on the issue, because of 'risks as a result of China's entry into the area.'[41] By 1964–5, after Indian and Pakistani leaders had separately visited the Soviet Union, Moscow's position on the dispute became neutral.

Pakistan's largest political gain in this period, however, was shown during the 1965 war when Moscow adopted an even-handed approach, called on both sides to exercise restraint and to work for a peaceful solution of the conflict.[42] Pakistan readily agreed to the Soviet offer to host a peace conference at Tashkent after the ceasefire, an offer that was also supported in the United States, but viewed with misgiving in India and in China. Although the Tashkent agreements achieved little, except to restore the status quo *ante bellum*, they appeared to demonstrate that Moscow had managed to establish itself as a major mediatory force in the region at Washington's and Beijing's expense.

The conference also served to further the Soviet relationship with Pakistan. Soviet–Pakistani trade began to increase as did Moscow's economic assistance: Pakistan was promised $51.7 million in 1965 and a further $84 million in 1966.[43] In June 1966, Pakistan started to press for Soviet arms. A few helicopters were delivered in 1967, and in May 1968, the two governments signed a formal arms assistance agreement, the quid pro quo being that Pakistan would request the United States to close the communications base at Bedaber.[44]

It was soon evident, however, that this arms transfer policy was beginning to endanger Moscow's relations with New Delhi. Soviet leaders therefore sought tangible evidence that its policy was worth the risk, particularly since India had responded by making overtures to China at the end of 1968 and early 1969.[45] What Moscow primarily wanted was Pakistan's support for its economic cooperation and collective security schemes, and real progress in Indo-Pakistani relations in the 'spirit of the Tashkent declaration'. Despite Soviet pressure, however, Pakistani leaders remained outspoken in their opposition to Soviet regional initiatives, often describing them as purely anti-Chinese alliances.[46] There was no sign either that tensions had decreased between the local states. In response, Moscow stopped all arms shipments to Pakistan in April 1970.

The maintenance of solid Sino-Pakistani ties, certain internal developments in India and Pakistan, and events outside the region, further encouraged Moscow to reassess the value of its even-handed approach in South Asia. The uncertain political trends in Pakistan in 1969 and 1970 contrasted unfavourably with those of stability in India where mid-term elections in September 1970 and general elections in March 1971 led to the consolidation of Mrs Gandhi's position. The coincidence of interest with India increased still further when war broke out on the Sino–Soviet border. In addition, the prospects of a Sino–American *rapprochement* (which Pakistan had facilitated) filled both New Delhi and Moscow with dismay. Not surprisingly, therefore, as the crisis in Pakistan erupted into civil war, Russia reaffirmed its commitment to India. In August 1971, the two allies signed the Indo–Soviet Treaty of Peace, Friendship and Cooperation that contained a defence clause pledging 'mutual consultations' and 'appropriate effective measures, to ensure peace and security for their countries' should either be attacked by a third party. Unlike in 1965, Moscow in 1971 sided unequivocally with India and blamed Pakistan for the outbreak of war.

The Soviet role in the conflict was an active one. Frequent consultations took place between Indian and Soviet defence chiefs, Moscow pledged that it would retaliate if China intervened, and responded to the US despatch of the *USS Enterprise* and a task force with naval deployments of its own. From August, it stepped up arms deliveries to India and on three occasions at the United Nations cast vetoes against cease-fire resolutions until the new leaders of Bangladesh had consolidated their control.[47] India had won and Pakistan had lost, but the war's outcome seemed to imply far more than this. As the International Institute for Strategic Studies report on the conflict confirmed:

> Nothing could have dramatized the new patterns more than the spectacle in December of the Soviet Union and the PRC vilifying

one another in the Security Council about the third Indo–Pakistani war while the United States shuffled on and off the sidelines.[48]

Prime Minister Bhutto, head of the truncated Pakistani state, might demonstrate his bitterness towards Moscow by referring to the Soviet ambasador at the United Nations as 'Tsar Malik',[49] but he knew only too well that he could not afford to sever relations with Moscow as a result of its role in the war. As Kosygin told Bhutto in 1972: 'If history were to repeat itself, we would again take the same position.'[50] Meanwhile, Moscow offered to help in the process of normalising relations among India, Pakistan and Bangladesh, postponed Pakistan's commercial debt repayment, resumed trading, and its aid projects.[51] Containment of China through the stabilisation of relations in South Asia remained Moscow's major objective.

The roots of hostility in South Asia clearly stem from partition but as the events of the 1960s and early 1970s indicate, Sino–Soviet penetration of the area has sustained and intensified the rivalry at the local level, thus reinforcing the structure of the South Asian security complex. Though it was American armaments that first disturbed the balance of power, Soviet and Chinese arms transfers have also fuelled an arms race. India's humiliation during its border war provided the major stimulus for New Delhi's massive armament and military modernisation programme, but the existence of the Sino–Soviet dispute allowed Moscow to offer India its most sophisticated weaponry and provided it an opportunity to consolidate and expand ties in all fields. Sino–Indian hostility stimulated close relations between Beijing and Islamabad, but the rift between Moscow and Beijing, with its clear geostrategic dimensions, increased the importance of those ties to China so that it too has established extensive military, political and economic links with Pakistan. The Sino–Soviet border conflict facilitated Sino–Soviet penetration of South Asia, but it was the Sino–Soviet dispute that enabled enduring and stable alignments to be forged with the local states.

One analyst has also suggested that India's large scale increase in its armed forces was partly responsible for the outbreak of the 1965 Indo-Pakistani war because Pakistan sought a military solution of the Kashmir problem before time ran out.[52] That war, and the US decision to terminate its weapons programme in the region – a decision that particularly affected Pakistan – increased the attractiveness of China as a source of political and military support. Though Pakistan's military connection with China has been less significant than Russia's links with India, nevertheless it has been useful in sustaining Pakistan as a military threat to India, and has complicated India's strategic picture.

Vociferous Chinese support for the Pakistani position on Kashmir from 1964 and Soviet support for India's stance in the 1950s has also helped to maintain intransigence over the problem. Without that support, it is possible that the Kashmir dispute would have lost its centrality. Moreover, if the South Asian states had been unreceptive to external offers of support, it is also likely that India's dominance in the region would have been confirmed, but based more on its domestic political stability, size, and international prestige, than on its military weight.

FLUIDITY IN ALIGNMENTS

Despite predictions to the contrary, within a relatively short period it became apparent that the 1971 war and its aftermath did not serve to draw India further towards the Soviet Union, neither did it result in India's overwhelming dominance of the sub-continent. Pakistan, though smaller in size and population, became more easily defensible. The war may have demonstrated some of the weaknesses in the Sino–Pakistani relationship, but China still remained a steady and faithful friend to Pakistan, and ties with Beijing still remained the cornerstone of Islamabad's policy. China continued as a major aid donor, converting $110 million in earlier credits to outright gifts, and it deferred payment for twenty years on the 1970 loan of $200 million. (Between 1964 and 1979, in fact, China offered in all some $620 million in assistance to Pakistan, representing 13 per cent of China's total aid programme.) Military assistance and cooperation also increased immediately after the war as four military delegations arrived in Beijing between the end of 1971 and January 1973.[53] China quickly made good the losses of military equipment Pakistan had sustained during the conflict. Indeed, its continuing arms supply has made it the backbone of Pakistan's military arsenal, supplying 75 per cent of its tank force and 65 per cent of the air force.[54] Furthermore, Washington soon announced that it would lift its embargo on arms sales to South Asia, thereby providing Pakistan with a dual source of supply denied to it in earlier years. That decision and the Sino–American *rapprochement* removed at least some of the complications in Islamabad's external relations.

China's decision to channel its aid to the neglected areas of the

North-West Frontier Province and Baluchistan showed its interest in maintaining the territorial integrity of the dismembered Pakistani state. In 1974 Beijing demonstrated continuing political support for Islamabad when it used one of its rare vetoes in the United Nations to prevent the entry of Bangladesh into that organisation, until the issue of the Pakistani prisoners of war had been resolved. Moreover, though the new government in Bangladesh, headed by Sheikh Mujibur Rahman, was inevitably close to Moscow and New Delhi, it was also relatively shortlived. Mujib's violent overthrow in August 1975 led to the installation of a regime committed to asserting its independence of India. To this end, it announced it would seek to establish diplomatic relations with China and Pakistan, a wish fully reciprocated by these two governments. China also quickly offered economic assistance to Dacca, and is reported to have helped with the training of the Bangladesh air force.[55]

As a result of these realignments in South Asia, India's regional supremacy has not proven to be overwhelming. Increased Soviet influence in New Delhi has also been difficult to detect from 1972 onwards. Moscow hoped, for example, to resume its role as mediator between India and Pakistan; India, however, preferred the bilateral approach.[56] Russia also continued to press Indian leaders to give their support to the concept of a collective security system in Asia. During Brezhnev's visit to New Delhi in November 1972, this proposal formed a major theme of the talks, but Mrs Ghandi continued to resist the idea, and even managed to avoid referring to it in the joint declaration produced at the conclusion of the visit.

Of major significance, China and India undertook a dramatic step forward in their relations in 1976. The slight thaw in 1969 and in 1970, when Mao sought out the Indian chargé d'affaires at the May Day function in Beijing, and the hints at normalisation of relations in 1973, were indications of shifts in mutual perceptions. But in 1976 India went further and offered to restore the level of its diplomatic relations to ambassadorial level, a gesture that China fully reciprocated three months later.[57] Trade relations were also re-established the following year, and on the political level, China gradually began to shift its position on Kashmir. Deng Xiaoping finally confirmed China's new stance in June 1980, when he declared the Kashmir dispute to be a bilateral problem between India and Pakistan which they should settle peacefully.[58]

The border problem has inevitably proved more difficult to resolve, despite five rounds of talks. During the Indian Foreign

Minister's visit to Beijing in 1979 and Huang Hua's return visit in June 1981 both sides agreed, however, that the unresolved boundary issue 'should not be an obstacle to the further development of bilateral relations'.[59] And in June 1980, Deng suggested that one way to settle the dispute was to agree to the present line of control (the Zhou Enlai proposal of 1962). China would then retain the Aksai Chin and India its claims in the North-East Frontier Agency, a solution that has not appealed to India so far.

Though it will prove difficult for a variety of reasons for India to accept this solution, including the likely reaction of Indian public opinion, the uncertain political climate following Mrs Ghandi's assassination, and pressure from the armed services,[60] it seems clear that neither side wants this issue to halt the normalisation process. Neither India's absorption of Sikkim as an associated state, its nuclear explosion in 1974, its recognition of the Heng Samrin government in 1980, or its public reaction to the Soviet intervention on Afghanistan, have prevented China from trying to improve relations. For India's part, it too has been willing to continue the dialogue despite the complications it introduces in Soviet–Indian relations, the strategic implications of the opening in 1978 of the Karakoram highway, its disapproval of China's attack on Vietnam, and increased Chinese economic and military support of Pakistan in the wake of the Afghan crisis. These events, and others of a similar kind, have served temporarily to raise the level of polemics between the Asian states, but have also provided the impetus to continue the efforts to reduce tensions.

Through an improvement in bilateral relations, India hopes to undermine the basis for the Sino–Pakistani relationship, lessen its own dependence on the Soviet Union (and be seen to be doing so), and reduce tensions on its borders with its largest neighbour. China supports the last two objectives and, like India, is engaged in a wider effort to reduce tensions and improve its external relations, even those with Moscow. Its serious attempts to modernise its economy require a peaceful international environment and a reduction in external commitments. In South Asia, it wants to encourage India's moves towards a more equidistant position between it and Moscow and between Moscow and Washington. China also approves of India's attempts to solve bilaterally issues in dispute with Pakistan, is pleased by India's continuing efforts to reject the Asian collective security system proposal, and appreciates Indian ministers' reiteration during contacts with Soviet leaders that New Delhi 'earnestly hopes' for the normalisation of relations with China.[61]

China has also had to come to terms with the incontrovertible evidence of India's growing military strength. Since 1964, India has spent approximately $30 billion on modernising its armed forces. It now has the third largest standing army, the fifth largest air force and eighth largest navy in the world. Its domestic arms industry is the biggest among non-Communist Third World countries, and it has the world's tenth largest industrial base and third largest supply of skilled and technical manpower to support this industrial and military growth.[62]

India's recent nuclear and space activities have also given China reason to reassess its strategic approach to the region. Beijing has inevitably been drawn into Indo–Pakistani nuclear rivalry, which has quickened since the 1974 explosion. Though there is little obvious benefit to Beijing in becoming involved in Indo–Pakistani nuclear politics, its long-standing friendship with Pakistan has required it to support Pakistan in the event of 'nuclear threat and nuclear blackmail'. Beijing may even have gone further since there are persistent but as yet unconfirmed reports that China has aided Pakistan's own nuclear programme, allegedly providing Pakistan with sensitive information about nuclear weapons' design, confirming that Islamabad's device would work, and testing a Pakistani weapon on Chinese territory.[63]

Yet despite this crucial new dimension to the Sino–Pakistani relationship, India's explosion and Pakistan's reaction has perhaps given China ever more reason to attempt to come to terms with New Delhi. Acquisition of a nuclear capability has succeeded in exacerbating India's ties with the USSR, reduced New Delhi's dependence on Moscow, and raised the future prospect of India matching China in the nuclear field as it now rivals China in conventional weapons.

Soviet fears regarding the normalisation of Sino–Indian relations and India's attempts to shift to a more non-aligned position in world politics have far outstripped its concerns about India's nuclear blast. Russia's propaganda themes and efforts to increase military and economic cooperation with India plainly illustrate this. Moscow has constantly warned New Delhi that Beijing has no intention of solving the border problem, or of stopping its aid to the Nagas and Mizos. It has also suggested to India that it should be wary of Sino–American attempts at encirclement, and has alleged that Washington and Beijing have identical interests in South Asia, which include using a rearmed Pakistan against India and the USSR.[64]

Moscow's approach has prompted New Delhi to remind Soviet leaders that they too have begun a dialogue with China and that 'all efforts' to ease tensions 'should be welcomed'. Indian leaders have also been required to reiterate, however, that any improvements of relations would not be 'at the expense of our time-tested friendship with other states'.[65]

Moscow attempted to demonstrate the vitality and benefit derived from the Indo–Soviet relationship by agreeing to sell Soviet crude oil to India in 1977 and by concluding one of its most generous arms agreements in 1980, an agreement which provided India with $1.63 billion in credit to purchase Soviet weapons over ten to fifteen years. The credit has been made repayable over fifteen years and carries a 2.5 per cent rate of interest. An exchange of visits in March 1982 and in June 1983 further consolidated military ties. On the first visit, the Minister of Defence, Dmitry Ustinov, led a 70–80 person delegation that included the air force and navy chiefs and several army generals. On the return visit, the Indian Defence Minister finalised a deal providing for the purchase of MiG 29s and licensed production of the Soviet T–72 tank and MiG 27 fighter bomber. This did not prevent India, however, from going ahead with an order for British Sea King helicopters and Sea Eagle missiles, or concluding licensing agreements for the West German Do–228 transport aircraft and the Jaguar.[66]

India's recent efforts and success in diversifying its sources of arms provides a major part of the explanation for Moscow's attempts to enhance the level of military cooperation with New Delhi. But another reason relates to Afghanistan and the Soviet need to gain understanding and support for its interventionist policy. In a number of respects, the events in Afghanistan have served to solidify the alignments in South Asia that we have been familiar with since the 1950s. Pakistan's strategic significance has increased dramatically in Washington's perception, since Soviet actions are depicted as being part of a plan to threaten Western access to the Gulf and Indian Ocean. The Reagan administration's $3.2 billion economic and military aid package (which includes the promise of 40 F–16 fighters) testifies to this new assessment of Pakistan's position. In reaction to Pakistan's enhanced relations with Beijing and Washington, Moscow regularly denounces Pakistan in language reminiscent of its threats of the 1950s. In May and June 1979, for example, the Soviet press accused Pakistani army chiefs of training Afghan rebels and noted that Pakistan was engaged 'in a risky game that may have dire consequences for it'.[67] In May 1981, *Izvestia* drew the parallel explicitly between the 1981 US aid programme to Pakistan and the

May 1960 U2 incident,[68] which resulted in Moscow's first nuclear threat to Pakistan.

More frequent exchanges of visits between Pakistani and Chinese leaders and statements of support for Pakistan's national independence and sovereignty have complicated both the Sino–Soviet and Sino–Indian dialogues. China has made a Soviet withdrawal from Afghanistan one of the central requirements for the normalisation of its relations with Moscow; and its fierce denunciations of USSR's policy, its limited support to the Afghan insurgents, and mutual strategic interest with Pakistan, add to the difficulty of reducing areas of conflict with Russia and India. In addition, the US F–16 arms agreement with Pakistan has enraged India but given China satisfaction, and the prospects of an alliance forming between Beijing, Washington and Islamabad as a result of the Soviet intervention, has become a major concern for New Delhi.

Nevertheless, China and India have been able to salvage something from the complexities introduced by the events in Afghanistan. Beijing has chosen to emphasise its close identity of view with New Delhi regarding Afghanistan, arguing that both fear further Soviet encroachment in South Asia. The inability of the USSR or India to make mention of the Afghan problem in any of the joint declarations issued after bilateral visits in the last few years, and instead to oppose 'all forms of outside interference in the region'[69] – a formulation that minimally satisfies all sides – testifies to a certain mutuality of interest between China and India that both find worth exploring.

The prospects of a genuine improvement in Sino–Indian relations and, above all, of a resolution of the border dispute, leads one to speculate what would have happened in the sub-continent if the 1962 war had not occurred. The cross-alignments would probably not have been made or made only loosely, and the massive armaments programmes that have been a feature of the last twenty years would not have seemed so imperative. India's nuclear programme may have been delayed or even not begun at all. It was only nine days after the Chinese test in 1964, for example, that the head of the nuclear establishment in India, Homi Bhabhan, argued that the only defence against nuclear attack 'appears to be the capability and threat of retaliation'.[7]

Similarly, if Sino–Soviet relations had not deteriorated to the level where each posed a major military and strategic threat to the other, then India and Pakistan would not have been so central to their rivalry and the pattern of alignments would not have been so rigidly adhered to. Moreover, the local powers would not so easily have found the extensive and enduring sources of external support that have fuelled their differences, and provided them with the confidence and capability to enter into hostilities with each other on two occasions in the last twenty years.

It follows, therefore, that if Sino–Soviet relations were to improve in the near future, it is possible that the divisions in South Asia would be softened. The softening effect is likely to be limited, however; limited by the maintenance of the basic reasons for hostility between India and Pakistan, limited by the memory of three major wars fought between them. Furthermore, Sino–Soviet *rapprochement* is not likely to be so complete that perceptions of strategic vulnerability associated with the South Asian border areas will diminish markedly. For this reason, the Sino–Soviet and South Asian security complexes are likely to remain interlinked, whatever improvements in relations take place at the local or super regional levels.

NOTES

1. *United States–Vietnam Relations, 1945–1967*. Study prepared by the Department of Defense. Washington, DC: US Government Printing Office, 1971, Book 1, pp. A56–58 (NSC 48/2, 30 December, 1949).
2. S.M. Burke, *Pakistan's Foreign Policy: An Historical Analysis* (London: Oxford University Press, 1973) p. 124.
3. Ibid., p. 255.
4. *American Foreign Policy*, 'Six Major Policy Issues: Address by the Secretary of State upon his return from a tour of the region', 1 June 1953, p. 2172.
5. *New Times*, no. 28, 12 July 1950, p. 20.
6. *Current Digest of the Soviet Press (CDSP)*, vol. v, no. 3, *Pravda*, 17 January 1953.
7. Robert C. Horn, *Soviet–Indian Relations: Issues and Influence* (New York: Praeger, 1982) p. 3, and Geoffrey Jukes, *The Soviet Union in Asia* (Sydney: Angus & Robertson, 1973) p. 106.
8. *CDSP*, vol. vii, no. 50, *Pravda* and *Izvestia*, 10 December 1955.
9. Horn, *Soviet–Indian Relations*, p. 8.
10. Charles B. McLane, *Soviet–Asian Relations*, vol. ii of Soviet–Third World Relations (New York: Columbia University Press, 1973); see chronological tables for Afghanistan, India and Pakistan.
11. *CDSP*, vol. vii, no. 33, *Izvestia*, 14 August 1955; *Ibid.*, vol. viii, no. 6, *Pravda* and *Izvestia*, 7 February 1956; M. A. Chaudri, 'Pakistan's Relations with the Soviet Union', *Asian Survey*, September 1966.
12. *CDSP*, vol. xi, no. 8, *Pravda* and *Izvestia*, 20 February 1959; McLane, *Soviet–Asian Relations*, p. 115.

13. *Summary of World Broadcasts, Far East (SWB, FE)*, BBC Monitoring Reports, no. 307, 23 November 1953; no. 311, 9 December 1953; *Survey of the China Mainland Press, (SCMP)*, US Consulate General, Hong Kong, nos 765, 767, 770, 776, March 1954.
14. Ibid., no. 869, 17 August 1954.
15. George McTurnan Kahin, *The Asian–African Conference, Bandung, Indonesia* (Ithaca: Cornell University Press, 1956) p. 28.
16. *Dawn*, 25 July 1957.
17. Stated by Lin Biao in his report to the Ninth Party Congress, April 1969.
18. J.D. Armstrong, 'Peking's Foreign Policy: Perceptions and Change', *Current Scene*, vol. xi, no. 6 (June 1973) pp. 3–8.
19. *New York Times*, 12 September 1969.
20. *CDSP*, vol. xxi, no. 23; *Pravda*, 8 June, no. 25; *Izvestia*, 20 June 1969.
21. Horn, *Soviet–Indian Relations*, pp. 36–42.
22. *National Assembly of Pakistan Debates*, Official Reports 2, 17 July 1963, p. 1665.
23. Allen S. Whiting and General Sheng Shih-ts'ai, *Sinkiang: Pawn or Pivot?* (East Lansing, Michigan: Michigan State University Press, 1958) p. 144; *SWB, FE*, no. 5022, 26 September 1975.
24. Allen S. Whiting, *The Chinese Calculus of Deterrence* (Ann Arbor: University of Michigan Press, 1975) esp. ch. 1.
25. Gerald Segal, *The Great Power Triangle* (London: Macmillan, 1982) p. 46.
26. John Gittings, *Survey of the Sino–Soviet Dispute* (London: Oxford University Press) pp. 160–61.
27. Segal, *Triangle*, p. 58; Horn, *Soviet–Indian Relations*, p. 8.
28. Neville Maxwell, *India's China War* (London: Jonathan Cape, 1970) p. 434.
29. *SCMP*, no. 3167, 24 February 1964.
30. Ibid., no. 4781, 10 November 1970.
31. Ibid., no. 3535, 5 September 1965.
32. Ibid., no. 3536, 8 September 1965.
33. See, for example, Anwar Syed, *China and Pakistan: Diplomacy of an Entente Cordiale* (London: Oxford University Press, 1974) and Golam W. Choudhury, *India, Pakistan, Bangladesh and the Major Powers: Politics of a Divided Subcontinent* (New York: The Free Press, 1975).
34. Syed, *China and Pakistan*, p. 149.
35. *SWB, FE*, 16 December 1971.
36. Jack Anderson, *The Anderson Papers* New York: Random House, 1973) pp. 260–61.
37. Ibid., p. 262.
38. *Dawn*, 4 February 1972.
39. Rajan Menon, 'India and the Soviet Union: A New Stage of Relations', *Asian Survey* (July 1978) p. 140.
40. Horn, *Soviet–Indian Relations*, p. 8.
41. G.W. Choudhury, 'Pakistan and the Communist World', *Pacific Community* (October 1974) p. 110.
42. *CDSP*, vol. xiii, no. 19; *Pravda* and *Izvestia*, 9 May 1965.
43. McLane, *Soviet–Asian Relations*, p. 122.

44. Choudhury, 'Pakistan and the Communist World', pp. 112–13.
45. *SWB,FE*, 1 January 1969, p. (i); Horn, *Soviet–Indian Relations*, p. 19.
46. *The Guardian*, 22 August 1969.
47. Horn, *Soviet–Indian Relations*, p. 72.
48. *Strategic Survey* (London: International Institute for Strategic Studies, 1971) p. 1.
49. *Observer Foreign News Service*, 14 January 1972.
50. Syed, *China and Pakistan*, p. 171.
51. *CDSP*, vol. xxiv, no. 10; *Pravda* and *Izvestia*, 19 March 1972.
52. Onkar Marwah, 'India's Military Power and Policy', in *Military Power and Policy in Asian States* (eds) O. Marwah, and J. D. Pollack (Boulder, Col.: Westview Press, 1980) p. 113.
53. *SCMP*, nos. 5075, 2 February 1972; 5130, 29 April 1972; 5239, 7 October 1972; 5301, 14 January 1973.
54. Yaacov Vertzberger 'The Political Economy of Sino–Pakistani Relations: Trade and Aid, 1963–1982', *Asian Survey* (May 1983) p. 647.
55. G.W. Choudhury, 'Post–Mao Policy in Asia', *Problems of Communism*, no. 26 (4), July 1977.
56. Horn, *Soviet–Indian Relations*, p. 79.
57. Bhabani Sen Gupta, 'South Asia and the Great Powers', William E. Griffith (ed.), *The World and the Great Power Triangle* (Cambridge, Mass.: MIT, 1975) p. 256; Horn, *Soviet–Indian Relations*, p. 123.
58. Horn, 'The Soviet Union in Sino–Indian Relations', *Orbis*, vol. 26, Winter 1983.
59. *Beijing Review*, no. 8, 23 February 1979, and no. 27, 6 July 1981.
60. Jerrold F. Elkin and Brian Fredericks, 'Sino–Indian Border Talks: The View From New Delhi', *Asian Survey* (October 1983) pp. 1135–6.
61. See, for example, *Beijing Review*, no. 7, 18 February 1980, no. 7, 15 February 1982, no. 41, 11 October 1982.
62. Marwah, 'India's Military Power', p. 101.
63. Vertzberger, 'Political Economy', p. 648.
64. Horn, *Soviet–Indian Relations*, p. 186.
65. *CDSP*, vol. xxxi, no. 24; *Pravda*, 12 June 1979; vol. xxxii, no. 7, 13 February 1980.
66. Horn, 'Afghanistan and the Soviet–Indian Influence Relationship', *Asian Survey* (March 1983) p. 256; *Strategic Survey*, 1983–84, p. 90; *SIPRI Yearbook*, 1984 (London: Taylor & Francis) p. 201.
67. *CDSP*, vol. xxxi, no. 15, *Pravda*, 10 April, no. 21, *Pravda*, 23 May 1979.
68. Ibid., vol. xxxiii, no. 18; *Izvestia*, 7 May 1981.
69. See, for example, *CDSP*, vol. xxxii, no. 50, *Pravda*, and *Izvestia*, 12 December 1980 on the conclusion of the Brezhnev visit.
70. Richard K. Betts, 'Incentives for Nuclear Weapons: India, Pakistan, Iran', *Asian Survey* (November 1979) p. 1056.

Part V
The Global Component of the Security Problem

From the discussions above, it is clear that the lines of cleavage within the South Asian complex act as a conduit for penetration into the region by outside powers. The basic mechanism here is the attempt by local states to use links with outside powers as a means of improving their position within the South Asian power structure. With the limited exceptions of Saudi Arabia's interest in Islam, and China's border dispute with India, none of the external states examined has much, if any, direct interest in the South Asian states, or the internal rivalries of the South Asian complex. The great powers get involved in South Asia much more in pursuit of their rivalries with each other, rather than because they have any concern for the outcome of rivalries within the region.

In this part, we focus on the globe-spanning complex of the superpowers as it relates to South Asia. As with the Sino–Soviet complex, we expect to find a pattern of alignment, and many of the questions which orient this discussion are similar to those set down in the introduction to Part IV. In the case of the Soviet–American complex, our principal interest is in the character of the penetration from higher to lower, and the impact of it on the security of the South Asian states. We examine the extent to which Soviet and American penetration of South Asia occurred as an offshoot of their rivalry, and the role played by the local states in drawing the superpowers in by request. We also look again at the impact of China on the superpower role in the sub-continent, and investigate the part played by the local pattern of insecurity in facilitating or resisting the superpower penetration. We try to evaluate how important South Asia is to the superpower rivalry, and therefore how deep and durable American and Soviet interests are in sustaining their current patterns of alignment. And we try to assess whether the superpower penetration has reinforced the local structure by intensifying its insecurities, or ameliorated it by imposing stability. We ask also whether American and Soviet involvement has favoured one side or the other in terms of its impact on the distribution of power.

8 The Superpower Global Complex and South Asia

ANITA INDER SINGH

INTRODUCTION

This chapter will examine the effects of the superpower global rivalry on the South Asian complex. It will address the following questions:

(1) Did the Soviet–American penetration of South Asia occur more as a result of the push generated by the global rivalry between them, or more as a consequence of demand pull from India and Pakistan? A related question is to what extent the superpower penetration has been conditioned by the subsequent and parallel pattern of Sino–Soviet penetration. Should the Sino–Soviet–American role in South Asia be considered as a single game or as two essentially separate, if overlapping, dynamics¿
(2) Has the impact of Soviet–American penetration been to exacerbate or ameliorate the dynamic of rivalry in South Asia and what are the relevant mechanisms involved?
(3) How important is the Soviet–American presence for maintaining the existing structure of the South Asian complex, and how does the superpower rivalry over Afghanistan affect the structure?
(4) How important is South Asia to Soviet–American rivalry? How deep and how durable are American and Soviet interests in maintaining their current pattern of intervention and is this pattern of intervention likely to change towards one of the greater or lesser intensity?

THE MOTIVES FOR SUPERPOWER ENGAGEMENT IN SOUTH ASIA

Did the Soviet–American penetration of South Asia occur more as a result of the push generated by the global rivalry between the superpowers or because of the demand from India and Pakistan? This question can be answered against the security perceptions of the superpowers, India and Pakistan, and the interaction between them.

The United States

We will begin by linking the main themes of US and Pakistani thinking which culminated in the Mutual Defence Assistance Treaty, and Pakistan's entry into SEATO and CENTO in May and October 1954 and August 1959, respectively. These were the first alliances between a superpower and an independent sub-continental state, and the ones to date on which some primary material is available. Contrary to some Indian assertions that the USA engineered the alliances to weaken India,[1] the evidence suggests that Pakistan soon after its creation, made the first approach to the US for military aid and a military alliance. Pakistan's motives lay in the circumstances that accompanied the partition of the sub-continent in August 1947. Pakistan symbolised the antithesis of the secular, independent India that the Congress had sought to achieve. From the moment of its birth on 14 August 1947, Pakistan waged a struggle for survival. Communal holocaust accompanied partition, and a disintegrating administration, inherited from the British, broke down completely. Statements against partition by Indian leaders, were contrued by Pakistan as Indian determination to destroy it. The mistrust of Indian intentions magnified by the military weakness of the new state, led Pakistan to search for military alliances and aid from the British and Americans soon after the British transferred power to India and Pakistan, such an alliance, had, in fact, been an essential element in the League's demand for Pakistan. As a political minority before August 1947, the League felt that Pakistan could not be achieved without British support. Having won their own state, they thought that the Western military umbrella was its only guarantee against Indian aggression.[2]

Pakistan first appealed to the US in October 1947 for $152 million over a five-year period to build up its armed forces.[3] How did the US respond? In seeking to contain the Soviet Union globally, US policymakers had their priorities. The strategic value to the US of non-Communist countries decided for it how American military aid would be disbursed, and priority consequently went to Western Europe. Under the heading 'Asia', a Policy Planning Staff report of February 1948 discussed only China and the Far East – no South Asian country was mentioned.[4] Neither India nor Pakistan met criteria determining American military priorities such as strategic location, terrain and economic ability to sustain a military build-up against the Soviets. Neither even had enough forces to meet its internal security requirements. In view of Pakistan's conflicts with both her neighbours, Afghanistan and India, it would be 'unwise' to give her military aid. Not surprisingly, then, the State–Army–Navy–Air Force Coordination Committee reacted warily to the Pakistani appeal for military aid in April 1949. It observed that Pakistan was thinking of the US as a primary source of military strength, and that a positive US response would involve 'virtual U.S. military responsibility for the new dominion'. No legal authority existed to grant such aid to Pakistan – probably because the US did not then have a military policy for South Asia. Furthermore, in December 1948, George Marshall was unimpressed by Liaqat Ali Khan's account of the importance of Pakistan in Western organised defence of the Middle East. At a time when the State Department thought the Middle East had 'no direct relation' to US security, Marshall responded that the US was 'already helping' Greece, Turkey and Iran, and there was a limit 'to what we could do'.[5] Summing up the official US attitude to Pakistan's requests for military aid in April 1950, a State Department Policy paper concluded that because of supply and priority problems, 'none of the significant requests had yet been complied with'.[6] As late as September 1950, George McGhee, then assistant Secretary of State, reflected the low priority given by the US to South Asia when he expressed the hope that the British would take the lead in bringing about agreement between India and Pakistan.[7] It is clear that until the end of 1950, the US did not respond to Pakistan's plea for close military ties because it considered such ties would be a liability and would contribute little to American global concerns.

The Communist takeover in China in September 1949, the outbreak of the Korean war in June 1950, signs of a Russo-Iranian *rapprochement* at the end of 1950, and the assassination of the Iranian Prime Minister, General Razmara Ali, on 7 March 1951, inspired an official reappraisal of US success in defending its security and in carrying out containment. Reports to the National Security Council by its chairman and the Chiefs of Staff observed that the US was on the defensive, especially to the Middle and Far East. The oil resources of the Middle East were of crucial importance to the US. Until now, the US had been involved in the setting up of NATO and had paid scant attention to the Middle East; the NSC did not produce any paper dealing with the Middle East as a strategic entity between 5 August 1948 and 27 December 1950.[8]

Breaking the official US 'silence' on the Middle East on 29 November 1950, McGhee asked the NSC to co-ordinate plans for Western organised defence of the Middle East. Both State Department officials and Chiefs of Staff pointed to the problems of organising defence in the region. Iraq had demanded the closure of British bases on its soil; the UK and Egypt had acrimonious differences over the Suez Canal; Arab–Israeli hostility meant the continuance of unsettled conditions in the area; and except for Greece and Turkey, most Middle East countries were professing neutrality between the Western and Soviet blocs.[9] On 2 May 1951, McGhee told the Chiefs that Turkey was joining NATO and wanted to raise twenty-five divisions. Iran also had to be defended. Whether Iran wished to be defended by the West he did not say. The money had to be found, and Congress would be approached for military aid to Iran. It is at this stage that McGhee asserted unequivocally:

> Pakistan wants to play a role in the Middle East ... They would do almost anything if the Kashmir problem could be settled. Liaquat is strongly on our side ... Pakistanis are good fighters and they can raise almost any number of men. Again ... there is an equipment problem to be solved. With Pakistan, the Middle East could be defended; without Pakistan, I don't see any way to defend the Middle East.[10]

McGhee's statement marked the shift in US thinking on Pakistan's military capability and role in the Middle East and the initial signs of a continuing American propensity to think of Pakistan as a South West Asian country. His assertion also showed how US global interests made it respond to Pakistan's plea for a military alliance. American officials were convinced by the summer of 1951 that Pakistan had given up searching for allies in South Asia and was now looking outside the region for alliances against the Chinese and Russians.[11] This is an amazing statement, for Pakistan had never sought alliances with India or Afghanistan since August 1947 – she had, on the contrary, wanted US help against them.

With most Middle East countries unable or reluctant to cooperate militarily with the US, American officials now saw Pakistan, whose regional hostility towards India had prompted her search for allies against possible Indian aggression, as an anti-Communist power, anxious for support against the Soviets and a willing and able partner in the Middle East. Indian neutrality was not the reason for the shift in the American perception of Pakistan.[12] Before 1950, the US was persuaded of the pro-Western bias of Indian non-alignment and had sought to build up India as the natural leader of a South Asian regional association.[13] But when US global needs dictated a military alliance with Pakistan all past doubts about Pakistani ability to contribute effectively to Middle East defence were cast aside, and by the end of 1951, the US had decided to include Pakistan in a Middle East Defence Organization.[14] The change from a Democratic to a Republican administration in 1953 did not alter the new US perspective on Pakistan. American global concerns also decided what would be offered to Pakistan. No guarantee was offered against an Indian attack, since the US had no conflict with India. What is of greater interest is that the US gave Pakistan no guarantee of support even against a Soviet attack on one of its SEATO partners. The US was not prepared to commit itself to 'specified actions' against Communist infiltration or subversion, but would only do what it considered 'necessary'. In 1955 the US Chiefs of Staff affirmed that the US could not guarantee the territorial integrity of any member nation of SEATO.[15]

Post-1954, the US never wrote off India, despite its continual annoyance at India's moral professions in world affairs. India received twice as much economic aid as Pakistan, leading to comment by some scholars that while the US decided what military aid should be given to its ally Pakistan, the Indians were free to spend as much as they liked on arms, and to acquire arms from any country. John F. Kennedy regarded the solution of its economic problems by India as the measure of success of democracy against Communist China, and described India as 'the hinge of fate' in Asia.[16] Global anti-Communism prompted the US and UK to offer India military help during her disastrous war with China in October 1962 – much to Pakistan's chagrin. But the US was neutral in the Indo–Pakistani conflict in 1965, which it regarded as a local dispute, and it cut off arms supplies to both parties. The global concern of the US in the mid-1960s was in Vietnam; this, and its then desire to contain China in Asia probably accounted for its concurrence with Soviet efforts to mediate between India and Pakistan at Tashkent in January 1966. So in both 1962 and 1965, the global perspective of the US did not converge with the regional perspective of Pakistan and left it indifferent to Pakistan's wishes, intervening once in favour of non-aligned India and abstaining from helping its ally in the second instance. However, in 1971, global considerations rendered the US indifferent to the brutal Pakistani repression in East Pakistan. At that time, Pakistan was secretly helping the US to reopen contacts with China, and the US regarded a diplomatic rapport with China as more important than developments in South Asia. Grafting China on to the international system would help the US balance of power position with the Soviet Union. The East Pakistan crisis was raised to the level of global geopolitics and the concern with China moulded all US reactions to it and the Indo–Pakistani war of December 1971. But having secured an opening to China through Pakistan, the Americans did nothing to prevent the break-up of Pakistan, although it was still a member of SEATO and CENTO. Involvement would probably have brought the US into conflict with India, which was not a global problem for it; and possibly into some confrontation with the USSR under the terms of the Indo-Soviet Treaty of August 1971, which the US would wish to avoid in an area not of vital importance to the Superpowers.[17]

Following the Soviet invasion of Afghanistan in December 1979, the primary US objective is to defend Gulf oil. With the perceived threat of a Soviet push towards the Gulf through Baluchistan, Pakistan represents 'high military value' to the US.[18] Its military is among the few in the Third World that can assimilate high technology weapons. The decision to supply Pakistan with F-16s is significant. There are few lines of Soviet advance towards the Gulf and these are subject to interdiction from the air. The F-16 can deliver air-to-surface weapons and it is the 'optimal aircraft' against a Soviet push towards the Gulf. The F-16 infrastructure introduced into Pakistan also establishes a baseline of inter-operability with the American and Saudi air forces, should circumstances require their support. Nevertheless, there seems little prospect of a military clash between the US and USSR as long as the Russians do not disturb the flow of oil from the Gulf. The commitment to Pakistan is minimal – a mere handshake compared with the embrace of the 1950s, which did not prevent the break-up of Pakistan either from internal causes or from external intervention.[19] On the whole then, the form of US penetration of South Asia has been determined by its global security perspectives.

Military ties, formal or informal, have been dependent on a convergence of interests between Pakistan and the US at a particular time.

The Soviet Union

Soviet aims in India were the same as in the rest of the Third World – to remove or to reduce Western influence. Stalin's ideological division of the world into socialist and non-socialist camps had little use for non-alignment, and Soviet encouragement of 'world revolutions' did not endear Russia to weak countries. To Stalin, India's decision to remain in the Commonwealth showed her servility to the West, which led to Nehru's retort that Soviet policy towards India was based on nonsensical premises, and as such, was bound to go wrong.[20] Khruschev's accession to power in September 1953 opened a new phase in Soviet foreign policy. The emphasis was now on widening the Soviet circle of influence by normalising relations with countries antagonised by Stalin and, in the Third World, to establish fresh contacts with newly independent countries and to steer them away from the Western bloc.[21]

The shift in Soviet foreign policy coincided with the formation of SEATO in October 1954. News of Pakistan's entry into SEATO struck a raw nerve of the Indian government. Nehru's dislike of alliances with superpowers stemmed from his belief that military alliances restricted the sovereignty of newly independent countries, and would bring India and Pakistan into the region of the cold war. The presence of a superpower in South Asia would hamper Indian efforts to judge each issue on its merits. India felt Pakistan's aim was to put pressure on her because of the Kashmir conflict and to encircle her with a ring of hostile alliances. Pakistan's military treaties with the US provided India with the pretext for reneging her offer of a plebiscite. The treaties also spurred on an Indian search for new sources of military strength to counter American influence on the sub-continent. The US–Pakistani treaties, unintentionally perhaps, opened new doors to Indo–Soviet ties. For SEATO also disturbed the Soviets. With seven countries, stretching from the US to the Phillipines, hostile to them, the advantages of Indian non-alignment were not lost on the Russians, and probably induced them to support the Indian stand on Kashmir after 1955. Kashmir is strategically situated from the Russian viewpoint, and it is easy to see why they would prefer its inclusion in a non-aligned India than in a hostile

Pakistan offering bases to the Americans. Close ties with India would also demonstrate peaceful Soviet intentions towards countries with different political systems. Thus a confluence of interests reflecting different political perspectives helped to forge closer Indo–Soviet ties.[22]

China's claims to Soviet territory and the first whiff of the Sino–Soviet ideological rift in 1959 introduced a new dimension to Soviet security perspectives. Since then, the containment of China in Asia has been a major goal of Soviet foreign policy. As South Asia has never been a high priority area for the USA, checking Chinese penetration of the area is a regional concern of the USSR. This accounts both for a ready Soviet response to the Indian quest for arms after its defeat by China in 1962, and the Soviet attempts to broaden their influence with both India and Pakistan during the mid-1960s. This resulted in Soviet mediation between the two countries at Tashkent in January 1966. By strengthening their ties with both India and Pakistan, the Soviets evidently hoped to stabilise their southern flank. But Indo–Pakistani differences over Kashmir and the Farrakka Barrage persisted after Tashkent, making it difficult for the Soviets to balance themselves between the two antagonists. Meanwhile, Soviet arms aid to India after its 1962 débâcle against the Chinese, and an American embargo on arms to Pakistan during its 1965 war with India, spurred Pakistani overtures to the Chinese for arms. This emphasised the deep-rooted hostility between India and Pakistan, for at that time, neither superpower had a strong position on the sub-continent, and both were hostile to China. Since 1965 Pakistani arms agreements with China have resulted in 75 per cent of its armed forces being equipped with Chinese weapons. Sinkiang and Azad Kashmir have been linked by roads through the Mintka and Karakoram passes in 1968 and 1971 respectively. June 1978 saw the completion of the 500 mile long Karakoram highway through the Khunjerab pass, the road having bridges designed to carry the weight of tanks. The roads have given China the potential to move forces and material rapidly and to outflank Indian forces in the region.[23] The Indians thus face the worrying prospect of a joint Sino–Pakistani attack on their northwestern and northeastern borders. The roads have strategic significance for the Soviets as well, with the possibility of the Chinese consolidating their position in Sinkiang with military facilities provided by Pakistan in Azad Kashmir.

With Nixon's election as president in 1968 appeared signs of an American withdrawal from Vietnam and of a Sino–US *rapproche-*

ment. The Russians were clearly concerned to stall Chinese attempts to fill the power vacuum created by the American disengagement from the Far East; the Indians were worried at the possibility of Chinese hegemony over South and South East Asia. US diplomatic moves towards China pushed India towards the USSR and away from the US after 1968.[24] For the Russians, Deputy Foreign Minister Nikolai Firyubin was emphatic in September 1968 that the threat of Chinese expansion was paramount in the Soviet perspective.[25] Soviet attempts to woo Pakistan with arms in late 1968 roused Indian ire; and the Kremlin recognised the unreliability of a political acquaintance which was moving closer to the Chinese militarily and economically.[26] The Soviets had to have at least one dependable friend in South Asia, and they calculated that the common Chinese threat offered opportunities for closer, more durable Indo–Soviet ties. Negotiations for an Indo–Soviet treaty began in 1969 and culminated in the agreement of August 1971, against the background of possible Chinese intervention on the sub-continent in favour of Pakistan.[27] Since 1971, attempts to ease India's problems with Pakistan, China and the US have not encountered any significant success, so the link with the USSR remains strong, especially in military matters. Indo–Soviet ties did not weaken even under an Indian government which sought to correct India's 'tilt' towards the Soviet Union between 1977–80. Important arms deals were made between 1972 and 1980, and in 1977, for the first time, the Soviets started selling precious crude oil to India. Thus the existence of a mutual antagonist in Asia – China – makes for the preservation of strong military ties between India and the USSR.

Conclusion

Penetration of South Asia by the superpowers has been prompted by their global security interest. Neither superpower has considered advantageous to itself the rivalry between India and Pakistan, and both have made some effort to ameliorate it: America in the 1950s and early 1960s and the Soviets during the mid-1960s. Paradoxically, superpower penetration has been facilitated by the existence of the Indo–Pakistani conflict. Pakistan appealed to the US soon after partition for military aid against India: the US responded only when it saw military ties with Pakistan advancing its global interests and with the exception of the movement of the Seventh Fleet in 1971 has not normally directed these ties against India. India and the Soviet Union began cultivating each other because their respective regional and global interests converged against SEATO in 1954. However, it was not the differences with the US and Pakistan but the post-1962 Chinese threat, and India's inability to get US arms aid, that induced India to accept Soviet military aid on a large scale. It is the common Soviet and Indian desire to contain China in Asia – for their different reasons – that has led to their present military ties. On the whole, superpower penetration of South Asia has occurred when their interests in pushing into the region have coincided with a demand pull by the local powers.

THE IMPACT OF THE SUPERPOWERS ON SOUTH ASIAN RIVALRY

Our next question is to what extent the Soviet–American penetration of South Asia has exacerbated or ameliorated the dynamic of rivalry in the region and through what mechanisms. Superpower penetration did not create the conflict between India and Pakistan; penetration has taken place largely when the security perspectives of the local and superpowers have appeared to converge. Superpower penetration of the sub-continent has taken economic, cultural, political and military forms. The non-military penetration does not seem to have aroused adverse comment on any side. The effect of non-military penetration has been limited in the sense that it has not affected or altered the internal structure of either country; nor has it contained unrest arising from their domestic problems. The continual internal crises – whatever their causes – heighten the political insecurity of the ruling élites or parties in India and Pakistan.[28] Military ties between one superpower and a local power have enlarged the extent of the local threat, domestic and external, in the security perspectives of both countries. Two factors help account for such fears. One is that military strength is the ultimate symbol and guarantor of a nation's security. The other is that a domestic crisis can be an invitation to external intervention which can aggravate domestic unrest; and in the case of Pakistan, led to its dismemberment in December 1971. Both factors partly explain Indian concern about the possible sinister role of a 'foreign hand' in recent developments in Punjab.[29]

Whether or not superpower military penetration has aggravated local rivalry can be partially appraised against what each local power gains by it. The US alliances with Pakistan did not end that country's search for security, which has been compounded by internal regional dissensions since partition. Nor did the alliances serve for Pakistan her primary purpose – security against disintegration through Indian intervention. Today, no formal alliance exists between the US and Pakistan, and the strengthening military ties since December 1979 offer Pakistan no formal guarantee of US support in the event of a Soviet threat to its territorial integrity.

It must be remembered that since 1966 it is China, not the US, which has been Pakistan's arms supplier. This is in spite of sizeable US–Pakistani arms deals since the Soviet invasion of Afghanistan. American military penetration of the sub-continent is therefore limited, in contrast with the Soviet military penetration of India since 1962. The Soviets have not merely sold arms to India but have also helped to build up its domestic arms industry, which is one of the largest in the Third World. Pakistan, on the other hand, possesses only an infant domestic arms industry which receives assistance mainly from the Chinese. Pakistan also faces the threat of a cut off of conventional US weapons if it manufactures nuclear weapons. But enforcement of the threat would further limit American military influence on the sub-continent, and many doubt that the US would really carry it out. For, in the words of an official of the Reagan administration, Pakistan 'is on our side'.[30]

American and Chinese arms assistance to Pakistan have made it possible for the Pakistani military to shore up its weak position in domestic politics as the only force capable of governing without a popular mandate. But it is important to remember that successive civilian political failures paved the way for military rule in Pakistan, and that American or Chinese military aid are not the *raison d'être* for the continued absence of a credible political alternative to military rule in Pakistan.[31] The Soviet invasion of Afghanistan has also enabled General Zia-ul Haq to continue martial law and has legitimised his political regime. At one stroke he has been able to transform his image in the West from that of a general who hanged his predecessor to a vital bulwark against Soviet expansionism.[32] Afghanistan will continue to be a crucial element in Pakistani politics. Zia's problem is that American and Chinese aid will not significantly lessen his domestic problems; yet there remains the danger that Soviet proximity to Pakistan may actually heighten regional tensions in a manner which could have adverse consequences for Pakistani unity. By enlarging the security threat faced by Pakistan the Soviet invasion of Afghanistan has also worsened relations between India and Pakistan. One reason has simply been their differing reactions to it. India has concentrated on attacking the Western response to the invasion rather than the invasion itself. India alleges – with justification – that in the past, American arms have always been used against it and not against the Soviets. India also claims that Pakistan will not be able to stand up to a Soviet attack.[33] India's fear of American supplies of the F-16 to Pakistan may arise partly from the demonstration of its capability in the successful Israeli raid on Iraq's Osirak nuclear plant, and the apprehension that Pakistan may stage a similar attack on Indian nuclear plants.[34] Thus US encouragement of the Sino–Pakistani axis can only enlarge India's perspective of the security threats she faces.

The USSR is the common factor linking the Indian and Pakistani quests for arms. The Russians not only pose a great immediate threat to Pakistan, but they also provide India with the bulk of its military supplies. This, in turn, can hardly persuade Pakistan of 'peaceful' or 'friendly' Indian intentions, especially when Indians have on occasion argued against US arms to Pakistan no matter how great the Soviet threat, and Mrs Gandhi's recent statement that Pakistan is a greater threat to India than the Soviet Union in Afghanistan.[35] The unresolved Kashmir conflict is, in a sense, no longer a local dispute. As a symbol of the hostility between them, it led Pakistan to approach the US for military ties, to counter which India showed a new interest in links with the Soviet Union in the mid-1950s. Thus the initiative by a local power took the Indo–Pakistani conflict outside the local structure. But South Asia is not a major bone of contention between the superpowers, who, as we have seen, have actually preferred for their own reasons, an easing of Indo–Pakistani tensions. In fact, it was the US refusal in the mid–1960s to give arms to both India and Pakistan on their terms against China and India respectively that pushed India into closer military links with the USSR and Pakistan into a military relationship with China. The need of each local power to counter the growing military strength of the other has contributed to a continuing spiral of arms acquisition by India and Pakistan. Essentially it is the persistence of their rivalry that militates against a slowing down of the sub-continental arms race. Until the Soviet invasion of Afghanistan in December 1979, Soviet–Chinese penetration in favour of India and Pakistan respectively exacerbated the conflict between the two local powers; since 1979, renewed American arms aid to Pakistan has aggravated it further.

THE SUPERPOWERS AND THE STRUCTURE OF THE SOUTH ASIAN COMPLEX

The importance of the superpower presence in maintaining the existing structure of the South Asian complex has to be assessed against the extent of their influence on the domestic and external concerns of India and Pakistan. The quickest answer could be made by raising the question: if neither superpower had direct interests in the region, would the structure of the complex change, and in what direction? As we have seen, Indo–Pakistani rivalry prevailed *before* either superpower evinced any interest in the sub-continent, and it was Pakistan which made the first move to a superpower to acquire weight in what it considered to be its inferior position in relation to India. If their animosity existed before the superpowers intervened, there is little reason to suppose it would end in the hypothetical event of a superpower disengagement from the sub-continent in the near future.

Non-intervention by the superpowers would probably have one of the following effects – more wars would either settle the regional balance, or aggravate regional instability, or consolidate Indian hegemony over the area. The question is again very theoretical, as the very perception of its inferior position would make Pakistan turn if not to a superpower, then to some other power. Pakistan would probably seek the assistance from China as it has done on a wide scale since the early 1960s. The China factor is in fact a major determinant of the South Asia complex. For Pakistan it is the regional power which can check Indian dominance of South Asia; on the Indian side it has motivated India's close ties with the USSR to balance the Sino–Pakistani collusion, and these ties are also a symbol of Soviet support against China. The China factor brought about the Indo–Soviet relationship after the mid-1960s and it is still the most important element in this link. Even without superpower intervention, China links Indo–Pakistani rivalry to the super-regional level, despite the fact that the possibility of war between India and Pakistan is more immediate than of a Sino–Indian war. That India has fought more wars with Pakistan than with China since 1947 does not diminish in her eyes the magnitude of the Chinese threat, any more than the absence of a direct Soviet–American conflict since 1945 has lessened for them the security risk posed by the other side. Again, as China modernises her armed forces, the extent of the Chinese threat as perceived by India will increase. Taken with China's military links with Pakistan, China will remain the primary factor in both maintaining, and exacerbating the local Indo–Pakistani rivalry.[36]

Yet the effect of the US perception of Pakistan as part of a South West Asian region should not be ignored. It has continually caused the US to underestimate the impact of its support for Pakistan on Indo–Pakistan relations – that is, the South Asian complex. Consequently, the US has effectively, though probably unintentionally, reinforced the structure of the complex by strengthening Pakistan. For Pakistan's major unresolved dispute is with India over Kashmir. Links between Sinkiang and Azad Kashmir have invested Kashmir with extraordinary strategical importance, especially with the Soviets in Afghanistan.

The Soviet occupation of Afghanistan has also preserved the local complex. It has pushed Pakistan to the US once again and has heightened US interest in South and South West Asia. Any Indian aspirations for regional hegemony have suffered a double setback by the interest of both superpowers in the area, so India can hardly welcome the Soviet invasion, despite her muted reactions to it. Pakistan's military links with the US oblige India to maintain her reluctant reliance on the Soviets, show her impotence to contend singlehanded with Sino–US support for Pakistan, and underline her inability to influence local events to her liking. As this probably suits Pakistan, Indo–Pakistani tensions have also been driven deeper, and will remain so, given the continuing, though differing interest of both superpowers in Afghanistan.

Superpower presence in the Indian Ocean has added another dimension to the local South Asian structure. The presence of the superpowers in the Ocean stems from their global security concerns. The US established its military presence in the Ocean during the early 1960s, before the Soviet invasion of Afghanistan, with the outstanding aim of having access to Gulf oil and protecting the transportation routes for its flow to the West.[37] Today, however, the Middle East supplies less than 20 per cent of the total world demand for oil, and it seems unlikely that oil provides the only *raison d'être* for American presence in the Indian Ocean. Probably it can be better explained by the perception of the international system by the superpowers as a bipolar zero sum game in which there can only be two main spheres of interest.[38] The Soviet presence in the Indian Ocean has arisen partly as a strategic response to the threat of American nuclear submarines to its southern flank, partly from its economic interests, and partly because the Indian Ocean is the direct link between its Far Eastern and Western Black sea coasts.[39] The US has bases in Somalia, Kenya, Oman, Egypt and Diego Garcia; the Soviets in South Yemen, Socotra, Aden, Ethiopia and the Seychelles.

How have the local South Asian powers reacted? Pakistan favours US naval presence in the Indian Ocean to counterpoise possible Indian dominance of the area. Support for its case comes from the absorption of small states by bigger ones if no balance of power conditions prevail – for example, the absorption of Tibet into China, and the current Soviet occupation of Afghanistan.[40] Outstanding in India's security perspective is the lesson of the American nuclear carrier *Enterprise* in the Bay of Bengal in 1971. Only a surge of Soviet naval deployment balanced the situation in India's favour. Today the US seventh fleet can reach Bombay in three days from Diego Garcia, at a time when the Americans are arming both India's antagonists. The Indians cannot discount the possibility of coercive diplomacy by the US in the future. India does not have the resources to challenge superpowers or to halt the arms race in the region, so Soviet naval deployments have a deterrent value against US capabilities in the Indian Ocean. With India and Pakistan unable to resolve their local disputes, superpower naval competition in the Indian Ocean will, on the whole, reinforce the existing structure of the South Asian complex.

SOUTH ASIA IN THE SUPERPOWER RIVALRY

The importance of South Asia to superpower rivalry and the durability of superpower interests in the sub-continent, can to some extent be gauged from the circumstances in which US and USSR chose to penetrate the area. Since the Second World War, South Asia has not been a primary potential battleground between them.

The United States

Pakistan was drawn into the network of US military alliances because of the role the US thought she could play in the Middle East defence, and not merely because of her proximity to the USSR. Indian non-alignment was not the motivating factor behind the US decision to forge military ties with Pakistan, and the US has resented India more because of the superior moral tone it has often adopted in international affairs than because of any strong political significance. In the mid-1960s, both superpowers wished to check Chinese influence in Asia, which led the US to concur with Soviet attempts to mediate between India and Pakistan in 1965–6. Then, in 1971, the US deployed the *Enterprise* in the Bay of Bengal largely to impress China; not to save Pakistan by getting involved in what it saw as a local conflict.[41] Today, its global rivalry with the USSR has persuaded the US to give financial and military assistance to Pakistan – there does not seem to be any great effort to get the Russians out of Afghanistan merely because of the threat they pose to Pakistan.

Pakistan gains strategic importance for the US to a very great extent because a pro-US government does not exist at the moment in Iran. Nor do most Middle East countries go along with American strategic plans for the region. But Pakistan's domestic tensions make her, in the eyes of many Americans, an unreliable ally at best.[42] In any case, US military aid for Pakistan is only envisaged at present until 1986. The US is opposed to the manufacture of nuclear weapons by Pakistan, while it assists the nuclear programme of a China which has already exploded the bomb. The latter is seen as a counterpoise to the USSR; while an 'Islamic' bomb could have an unsettling effect on an already unstable but strategically important Middle East. The point is that for themselves, neither Pakistan nor the South Asian region are of great interest to the US in its rivalry with the USSR.[43]

Economically, too, neither India nor Pakistan are very important to the US. Protected industries and standards of living limit their markets for American goods, while the American market is India's largest.[44] This obviously makes the US more important to India economically than vice versa.

The US *does* want a say in any settlement Pakistan might make with the Soviets over Afghanistan. But this has more to do with tactics against the USSR than concern for Pakistan's interests. US–Pakistan ties could also be affected by a change of administration. Carter upgraded India; Reagan sees Pakistan playing a role in American containment of the Soviets and has downgraded India.[45] Whether Pakistan will make a settlement with the Soviets even without American support is doubtful. But its dependence on the US after 1979 might stall any hopes of an agreement – as reportedly in the summer of 1983.[46] At the same time, any renewed American support for Afghan guerrillas will hamper the possibilities of a settlement, and will maintain superpower interest in Afghanistan.

Given the existing Indo–Pakistan tensions, an intensification of superpower rivalry could also exacerbate the local conflict.

India is irrelevant in US policymaking, especially after its ambivalent stand on the Soviet invasion of Afghanistan.[47] She appears pro-Soviet to the Americans, who ignore her sharp reactions to their arms deal with Pakistan. The US has annoyed India, in turn, by abstaining on a vote on a $5.8 billion IMF loan to India, and has blocked aid for technological plants and nuclear power stations on the ground that sensitive technology should be restricted. Here it is the cold war mentality that counts. The US world view also accounts for its espousal of permanent Chinese membership of the International Atomic Energy Agency.[48]

The recent US concern at rising Indo–Pakistani tension, even talk of war, probably reflects American worry that it might have to intervene in a local conflict. If the US came down on the side of Pakistan, India could be pushed decisively into the Soviet camp, and South Asia could become a new theatre of superpower conflict, adding to US 'responsibilities'. Non-intervention on the other hand, could result in the defeat of its ally Pakistan, which is presently the key to any direct application of Western influence in Afghanistan,[49] and, most importantly, would symbolise a victory for the Soviets. At best, neutrality or non-intervention would earn the US the opprobrium of both sides. At a time when the US seems reluctant to intervene in South Asia even when invited,[50] an Indo–Pakistani war is something that it would rather see avoided.

Not in itself an area of vital strategical interest to the US, South Asia, especially Pakistan, assumed importance in the US perspective out of cold war considerations. The US interest in South Asia has been relatively high in the early 1950s and post-1979; it was low in the 1960s when the Americans were involved in Vietnam. In any case, Pakistan's significance to the US originates from the American propensity to think of her as a *Middle Eastern* power which can be strengthened against the Russians; in South Asia itself, the US has tried to maintain a balance of power between India and Pakistan. Given that the Middle East is of vital importance to the US, would the Americans respond more strongly if Pakistan's territorial integrity was threatened directly by the Russians? The Pakistanis themselves are uncertain, for US strategy will be determined, in the last analysis, by the perceptions and conceptions of the 'national interest' of the numerous interests at play in the US at a particular time.[51]

The Soviet Union

Like the US, the Soviet interest in South Asia stems primarily from the spheres-of-influence rivalry between them. It has been observed that South Asia 'has been second only to the Middle East as the site of the largest sustained Soviet attention and investment in the Third World'.[52] It appears then, that the Soviets consider their interest in the region quite durable. Of this, their attitude to their relationship with India bears ample proof. Soviet 'investment' in India has taken various forms – cultural, economic, political and military. The cultural has probably been the least influential for, in the words of a leading Indian Sovietologist, 'even Indian communists don't quote Soviet intellectuals'![53] Economic influence has shown better results for both sides. The USSR has financed 30 per cent of India's steel capacity, 70 per cent of its oil extraction facilities, 30 per cent of its oil refining capacity, 20 per cent of its power generating capacity and 80 per cent of its metallurgical engineering production. Such substantial aid to the Indian public sector has reduced India's economic dependence on the West. The terms of Soviet loans have been generous, with an interest of 2.5 per cent to be paid over 12 years in rupees, while Western offers carry interest of 4 to 6.5 per cent, to be paid in four to five years in convertible currency. Indo–Soviet trade has grown by leaps and bounds. In 1953, total trade between the two countries was worth $1.6 million – 'a total surpassed even in the closing days of Tsarist Russia.'[54] Today, the USSR is India's fastest expanding market, with trade between the two countries worth more than $2 billion, and expected to double by the end of the present decade.[55]

The sale of Soviet crude oil to India since 1977 has economic and political significance. The Soviets started selling crude to India against rupee payments. 'If there was one step the Soviets could take to demonstrate to Indian and to third parties the importance of Moscow's attitude to its relationship with New Delhi the provision of crude oil was it.' The supply of Soviet oil has been very important since the outbreak of the Iran–Iraq war – those two countries formerly met 70 per cent of Indian oil imports. Moscow cut off oil supplies to West Europe to increase supplies to India and other Third World countries.[56] It is significant that 'the single most important point' made by a visiting Soviet delegation to New Delhi in March 1983 was that if a war in the Gulf region jeopardised Indian oil supplies, India could count on the USSR to rush to meet oil requirements.[57] Soviet oil could become very important in the event of an Indo–Pakistani war, especially if Pakistan's Middle East friends barred oil supplies to India. With India producing only 40 per cent of her oil requirements, Soviet oil is a significant and durable element of the Indo–Soviet relationship, and which could even become India's lifeline in a war.

Oil is an important means by which the Soviets can keep alive a trade relationship which is running into some difficulties. India considers the USSR an unreliable trading partner, and simultaneously looks to the West and Japan for high technology which the Russians cannot provide. If India becomes more dependent on the West for technology, her dependence on the USSR will lessen, weakening a lever of economic – and political – Soviet influence over India.[58] What has the Soviet Union gained politically from the Indo–Soviet treaty of August 1971? Essentially, the treaty has formalised a legitimate regional role for the Soviets in South Asia and has set the pace for similar treaties with other Third World countries.[59] The treaty also formalised the anti-China link between India and the USSR. It has not, however, committed the USSR to any specific course of action. Its immediate effect in 1971 was as a psychological deterrent against Chinese intervention on behalf of Pakistan in the East Bengal situation. A Pakistan armed by China, with whom the Americans were then seeking a *rapprochement*, was defeated by an India supplied with Soviet arms. What was significant about this for the Soviets? To quote a Soviet diplomat at the UN, 'This is the first time in history that the United States and China have been defeated together'![60]

Nevertheless, Moscow has secured only limited gains from the treaty. In 1971 it did not anticipate India's use of the treaty to dismember Pakistan and to become the dominant power on the sub-continent. Nor has Moscow obtained facilities for its Indian Ocean flotilla, or for Brezhnev's Asian security plan which was mooted in June 1969. It has not been able to dissuade Indian attempts to normalise relations with the US and China; nor was it invited to mediate in 1972 between India and Pakistan as at Tashkent in 1966. The Soviet Union has obtained India's 'understanding' on issues that have little direct relevance to India such as Czechoslovakia and SALT. India's ambivalent stand on Afghanistan has cost it some influence over Third World countries, who have not endorsed the Soviet occupation of Afghanistan.[61] But the USSR has not secured any privileged strategic foothold in India.[62]

With limits to the economic and political levers of Soviet political influence over India, what does the Indo–Soviet military relationship portend? During the 1950s and early 1960s, Soviet military aid to India lagged behind economic aid and trade programmes. Only after 1963 have big arms deals been made between the two countries. But it is the 1983 Soviet decision to sell India the MiG–29 that may illustrate best the Soviet anticipation of the durability of its ties with India. The desire for sophisticated technology has induced India to turn to the West, not least to counter the advantages Pakistan reaps from possessing the F–16. The Russians have been clearly unhappy at Indian attempts to diversify the sources of arms. Indian officials were reportedly taken by 'considerable surprise' at the Soviet MiG–29 offer, for the aircraft is not even in service with the Soviet airforce as yet. India will also receive military technology at par with that given to Warsaw Pact countries, and more than 60 per cent of the new weapons and technology for the Indian airforce and navy will come from the USSR.[63]

Many Indians fear that the MiG–29 offer indicates the Soviet desire to maintain Indian dependence on Soviet arms, and that India's ambition to maintain regional dominance might be achieved with a very high political price tag.[64] Arms and oil are the two spheres in which the USSR can assure a mutual durability of Indian and Soviet interests. For the USSR, this is important for its position in Asia. The ASEAN countries have little love for the Russians; the Chinese threat looms over South and South East Asia; Bangladesh expelled Soviet diplomats recently, and there have recently been unconfirmed reports of offers of bases by Bangladesh and Sri Lanka to the US, while Pakistan's links with China and the US will continue to be a sore point with the Soviets. In most of Asia, and certainly in South Asia, India provides the only possibility of a break in the American circle of containment, and it is easy to discern why the Soviets are eager to preserve their relationship with a non-aligned but regionally dominant country, and are eager to aid the enhancement of that dominance. This contrasts with the American policy of trying to maintain a balance of power situation on the sub-continent, much to India's annoyance.[65]

The Soviet occupation of Afghanistan points to a durable Soviet interest in South Asia. There is no sign of an early or unconditional withdrawal from that country. The Russians do acquire gains from Afghanistan – cheap Afghan gas at a price well below world level; air bases at Shindad on the Iranian frontier, which borders the Baluch areas of Iran and Pakistan, and offers the Soviets possibilities of threatening to stir up domestic conflict as a potential without actually invading those countries. It thus keeps its southern flank secure.

What does a prolonged Soviet presence in Afghanistan augur for the South Asian complex? The enhancement of influence of either superpower stands in the way of Indian aspirations to regional hegemony and goes against the grain of their desire to keep superpowers out of South Asia. Since the mid-1960s, India has needed Soviet help to counter the Chinese and Sino–Pakistani axis regionally. The Russian refusal to withdraw from Afghanistan will draw China and Pakistan – and possibly the US – still closer to each other, so that Indo–Soviet ties are likely to remain durable, despite continual Indian efforts to avoid enmeshment in any Soviet apron strings. Even if a decisive Indian swing to the West occurs for economic reasons, it seems unlikely that it would markedly affect either the basic structure of the South Asian complex, or the pattern of superpower rivalry that has been grafted on to it because of the coincidence of interest of the local and superpowers.

CONCLUSIONS

Superpower penetration of South Asia has been motivated by the global rivalry between the US and the USSR; it has resulted in alliances with local powers when their has been a convergence of interests between the super and local powers. Penetration has therefore occurred because of simultaneous push from the superpowers and pull from the local powers.

Since the 1960s, Chinese penetration of the local South Asian complex has exacerbated the dynamic of the local rivalry. Sino–Soviet penetration of the sub-continent stems from their regional rivalry in Asia: China does not challenge the USSR globally. The Soviet–American penetration has its origins in their global animosity; the US wishes to exploit Sino–Soviet rivalry in *Asia* to contain the Soviets in South and South East Asia. The Sino–Soviet regional rivalry in Asia and the Soviet–American global conflict are therefore two separate, if parallel dynamics.

Penetration by the superpowers and China has followed the pattern of the local rivalry: military links between the US and Pakistan and China on the one hand, the USSR and India on the other, have exacerbated the dynamic of the local rivalry.

Superpower penetration of the Indian Ocean and Afghanistan also originates in their global rivalry, and not from a convergence of interest between the super and local powers. Because of their limited military resources, the local powers cannot challenge superpower penetration. Their own local rivalry induces them to lean on one or the other superpower to preserve their individual positions against their local opponent. If the local disputes remain unresolved, then superpower penetration in the Indian Ocean and Afghanistan would probably reinforce this pattern. An intensification of superpower rivalry in South Asia could then lead to a deepening of the local conflict as well.

NOTES

1. For example, see I. Gandhi, 'India and the US', *Foreign Affairs,* vol. 51, no. 1972, p. 50, B. R. Nayar, 'Treat India Seriously', *Foreign Policy,* no. 18 (1975) pp. 133–54.
2. IBCM paper IB(47)89, 27 May 1947, *TOP,* vol. 10, pp. 1004–8; Ismay's report, 8 October 1947, presented to Cabinet Commonwealth Affairs Committee on 14 October 1947; note by A. Carter, 8 October 1947; memorandum by Secretary of State for Commonwealth Relations, 'Aid to Pakistan-Supply of Military Stores', CAB 134/54. See also SANACC report, 19 April 1949, *FR,* 1949, vol. 6, pp. 13–28.
3. SANACC report, 19 April 1949.
4. PPS report, 24 February 1948, *FR,* 1948, vol. 1, pt 2, pp. 510–29.
5. SANACC report and Secretary of State to Acting Secretary of State, 29 October 1948, *FR,* 1948, vol. 5, pt 1, pp. 435–6.
6. State Department policy paper on Pakistan, 3 April 1950, *FR,* 1950, vol. 5, p. 491.
7. Record of US–UK discussions on 18 September 1950, *FR,* 1950, vol. 5, p. 198.
8. Editorial note, *FR,* vol. 5, pp. 1–2.
9. Memorandum from McGhee to Secretary of State, 27 December 1950, ibid., pp. 4–11.
10. State Department paper around 27 December 1950, ibid., pp. 11–14 and State Department draft discussion with Joint Chiefs of Staff, 2 May 1951, ibid., p. 120.
11. State Department Policy statement on Pakistan, 1 July 1951, *FR,* 1951, vol. 6, pp. 2208.
12. An NSC staff paper of 1 June 1949 listed both India and Pakistan as 'probable neutrals' in a world war. *FR,* 1949, vol. 1, p. 342. See also Appendix B to SANACC report, 19 April 1949, vol. 6, p. 28, and State Department regional policy statement on South Asia, 9 October 1950, *FR,* 1950, vol. 5, pp. 247–8.
13. Ibid.

14. M. Ayoob, 'India as a Factor in Sino–Pakistani Relations', *International Studies,* January 1968, p. 283; States Department policy statement on Pakistan, 1 July 1951, *FR,* 1951, vol. 6, p. 2208 and Record of conversation between Foreign Office and McGhee, 3 April 1951, FO 371/92875.
15. M. Venkataramani, *The American Role in Pakistan* (New Delhi, 1982) pp. 351 ff.
16. S. M. Burke, *Pakistan's Foreign Policy* (Oxford, 1973) pp. 323, 328–30.
17. This account is based on, R. Jackson, 'The Great Powers and the Indian Subcontinent', and W. J. Barnds, 'India and America at Odds', *International Affairs,* vol. 49, no. 3 (1973) pp. 35–50 and 371–84 respectively; C. van Hollen, 'The Tilt Policy Revisited: Nixon–Kissinger Geopolitics and South Asia', *Asian Survey,* vol. 20, no. 4 (1980) pp. 339–61; R. G. C. Thomas, 'Security Relationships in South Asia: Differences in the Indian and American Perspectives', 'Between the Stools? US Policy Towards Pakistan During the Carter Administration', *Asian Survey,* vol. 22, no. 10, pp. 959–77.
18. R. G. Wirsing and J. M. Roherty, 'The United States and Pakistan', *International Affairs,* vol. 58, no. 4, pp. 593–4.
19. M. Ali, 'Soviet–Pakistan Ties Since the Afghanistan Crisis', *Asian Survey,* vol. 23, no. 9, pp. 1030–2. See also B. Sen Gupta, 'Scrambled Strategic View', *India Today,* 15 November 1982, p. 68.
20. J. A. Naik, *Soviet Policy Towards India* (Delhi, 1970) pp. 30 ff, and S. Gopal, *Jawaharlal Nehru: A Biography* vol. 2, *1947–1956* (London, 1979) pp. 44–5.
21. Naik, *Soviet Policy,* pp. 68 ff.
22. Ibid., pp. 90–1, and Gopal, *Nehru,* pp. 186–7.
23. This paragraph is based on B. Sen Gupta, *The Fulcrum of Asia* (New York, 1970), and J. Vertzberger, 'The Political Economy of Sino–Pakistani Relations', *Asian Survey,* vol. 23, no. 5 (1953) pp. 647–8.
24. Mrs Gandhi's interview with C. L.Sulzberger, cited in R. Horn, *Soviet–Indian Relations* (New York, 1982) p. 34.
25. Ibid., p. 27.
26. Vertzberger, 'Political Economy of Sino–Pakistani Relations'.
27. For background of Treaty, see Horn, *Soviet Indian Relations; R. H. Donaldson, Soviet Policy Towards India* (Cambridge, 1974) ch. 6. See also J. P. Chiddick, 'Indo-Soviet Relations, 1966–1971', *Millennium,* vol. 3, no. 1 (1974) pp. 17–36.
28. On Pakistan's domestic politics, see K. B. Sayeed, *Politics in Pakistan* (New York, 1980) pp. 187–8; L. Ziring, 'Pakistan's Nationalities Dilemma: Domestic and International Relations', in Ziring (ed.), *The Subcontinent in World Politics* (New York, 1978) pp. 89 ff, and A. Saikal, 'The Pakistan Unrest and the Afghanistan Problem', *World Today,* March 1984, pp. 102–7.
29. Interview with President Zia-ul Haq, *India Today,* 15 July 1984, pp. 35–7, and editorial, 'The Foreign Hand Syndrome', *India Today,* 15 August 1984, p. 4; and ibid., 15 December 1984, pp. 70–1; R. Kothari, 'A Fragmented Nation', *Seminar,* January 1983, pp. 24–9.
30. *Newsweek,* 9 July 1984, p. 39.

31. Saikal, 'The Pakistan Unrest', *World Today*, March 1984, pp. 102–7 and J. Elliott, 'Waiting for General Zia', *Financial Times*, 2 May 1984.
32. *Newsweek*, 9 July 1984, p. 39.
33. *Far Eastern Economic Review*, 25 September 1981.
34. Ibid., 31 May 1984, p. 30.
35. R. Litwak, 'The Soviet Union in India's Security Perspective', in R. Litwak et al., *India and the Great Powers*, International Institute of Strategic Studies (1984) p. 112.
36. S. Nihal Singh, 'The Convenience Factor', *Illustrated Weekly of India*, 30 October 1983, and 'Why India Goes to Moscow for Arms', *Asian Survey*, vol 25, no 7 (1984) p. 719; Inder Malhotra, 'India's Defence Problems', *Times of India*, 19 January 1984.
37. On Indian Ocean politics see among others, A. J. Cottrell and R. M. Burrell (eds), The Indian Ocean: *Its Political, Economic and Military Importance* (New York, 1972); K. R. Singh, *Politics of the Indian Ocean* (Delhi, 1974); A. J. Cottrell and associates, *Sea Power and Strategy in the Indian Ocean* (London, 1981); L. W. Bowman and I. Clark, *The Indian Ocean in Global Politics* (Boulder, 1981); R. B. Rais, 'An Appraisal of US Strategy in the Indian Ocean', *Asian Survey*, vol. 23, no. 9 (1983) pp. 1043–51; C. Kumar, 'The Indian Ocean: Arc of Crisis or Zone of Peace?', *International Affairs,* vol. 60, no. 2 (1984) pp. 23–46.
38. K. Subrahmanyam, 'An Indian Perspective on International Security', in D. H. McMillen (ed.), *Asian Perspectives on International Security* (Macmillan, 1984) p. 154.
39. R. H. Donaldson, 'The USSR and the Indian Ocean', in Ziring (ed.) *The Subcontinent in World Politics*, pp. 168–95.
40. The point is made by R. G. C. Thomas, 'Security Relationships in Southern Asia', *Asian Survey*, vol. 21, no. 7 (1981) pp. 697–8.
41. P. I. Cheema, 'Pakistani Perspectives on International Security', in McMillen (ed.), *Asian Perspectives on International Security*, p. 147.
42. S. Harrison, 'For Pakistan, Effective Aid', *International Herald Tribune*, 9 January 1981; and C. Van Hollen, 'Leaning on Pakistan', *Foreign Policy*, no. 38 (1980) p. 50.
43. For recent reports of US Pakistan nuclear controversy, see for example, J. N. Parimoo, 'President Zia's Visit to US', *Times of India*, 21 December 1982; *International Herald Tribune*, 22 June 1984 and editorial, 'Chinese Nuclear Pact', 3 July 1984; *South*, July 1984; *South*, July 1984 and *Far Eastern Economic Review*, 2 August 1984.
45. See for example, Wirsing and Roherty, 'The United States and Pakistan', *International Affairs*, vol. 58, no. 4 (1982) p. 589 and J. S. Mehta, 'A Neutral Solution', *Foreign Policy,* No. 47 (1982) pp. 140, 147.
46. *India Today*, 31 July 1984, p. 49.
47. On Indo–US differences, see among others, I. Badhwar, 'Indo–Pakistan Tangle', *Indian Today*, 15 May 1984, p. 67; *International Herald Tribune*, 16 November 1981; R. G. C. Thomas, 'Security Perspectives in Southern Asia', *Asian Survey*, vol. 21, no. 7, pp. 689–709.
48. This account is based on *Financial Times,* 29 October 1984 and Hari Jai Singh, 'India and IAEA', *Indian Express,* 11 April 1984. In accordance with a compromise formula, both India and China are among the 'globally advanced nuclear nations' on the IAEA board.

49. H. S. Bradsher, 'Afghanistan', *Washington Quarterly,* vol. 7, no. 3, 1984, p. 46. See also W. E. Griffith, 'The USSR and Pakistan', *Problems of Communism,* January–February 1982, pp. 41, 44; J. Vertzberger, 'Afghanistan in China's Policy', ibid., May– June 1982, pp. 19–23; and P. I. Cheema, 'The Afghanistan Crisis and Pakistan's Security Dilemma', *Asian Survey,* vol. 23, no. 3, March 1984, p. 52.
50. This refers to reports of the US declining Sri Lanka's appeal for military assistance to help control Tamil unrest. *The Economist,* 15 September 1984, p. 52.
51. See Z. Khalilzad, Afghanistan and International Security'; P. I. Cheema, 'Pakistani Perspectives', in McMillen (ed.), *Asian Perspectives on International Security,* pp. 126–49.
52. J. L. Nogee and R. H. Donaldson, *Soviet Foreign Policy Since World War II* (New York, 1981) p. 161.
53. B. Sen Gupta, quoted in *Far Eastern Economic Review,* 16 August 1984, p. 28.
54. R. C. Horn, *Soviet–Indian Relations,* pp. 7–8 and 169–70; P. S. Ghosh and R. Panda, 'Domestic Support for Mrs. Ghandi's Afghan Policy: the Soviet Factor in Indian Politics', *Asian Survey,* vol. 23, no. 3, pp. 268–9.
55. *Financial Times,* 8 August 1984.
56. R. C. Horn, *Soviet–Indian Relations,* p. 139.
57. B. Sen Gupta, 'For Better or For Worse', *India Today,* 31 March 1984, p. 44.
58. The Soviets would naturally like to avoid this. See *Financial Times* report, 'Western electronics in Indo–Soviet barter agreement', 29 October 1984.
59. Z. Imam, 'Soviet Treaties with Third World Countries', *Soviet Studies,* vol. 35 no. 1, p. 60. Also A. Z. Rubinstein, *Soviet and Chinese Influence in the Third World* (New York, 1975) pp. 42–4.
60. *The Hindu,* 13 December 1971.
61. A. Z. Rubinstein, *Soviet Foreign Policy Since World War II,* (Cambridge, 1981) p. 223.
62. Ibid.
63. *India Today,* 15 August 1983, pp. 40–2; and B. Sen Gupta, 'For Better or For Worse', p. 43.
64. *India Today,* 31 August 1984, p. 50.
65. H. W. Wriggins, 'The Range and Scope of US Interests in South Asia', in Ziring (ed.), *The Subcontinent in World Politics,* p. 219.

Part VI
Conclusions

9 The Future of the South Asian Security Complex

BARRY BUZAN AND GOWHER RIZVI

In conclusion we return to the framework established in the Introduction. We need to reintegrate the four levels of analysis so as to draw into a single perspective the principal themes and observations from our studies. We also want to use the preceding analyses as a basis for speculating about the possible futures of the South Asian security complex. For both these tasks, our vehicle is the four models of change outlined earlier: the status quo, the internal transformation, the external transformation, and the overlay. How do the major patterns of continuity and change in South Asia fit into these models?

STATUS QUO

The case for a status quo future in South Asia logically requires the exclusion of all three alternative models. It has three principal foundations: first, that many of the main elements and relationships which define the complex appear to be stable, second, that the areas in which change looks most probable do not seem likely to undermine the existing essential structure of the complex, and third, that the kinds of changes necessary to alter the status quo do not appear probable, at least in the near future. The case for the status quo looks strong in the short, and even in the medium, term.

As argued in Chapters 2 to 5, most of the internal components of the South Asian complex appear durable. India, despite its many ethnic and religious divisions, and its ingrained political and economic inefficiencies, is unlikely either to fall apart as a result of internal tensions, or to allow external actors to dismember it. New Delhi's internal political troubles will continue to absorb much of the country's political energy, and long-term stability will depend on its ability to resolve satisfactorily the difficult question of centre-state relations. But these problems are highly unlikely to overwhelm a state which enjoys both a popular legitimacy, and command over substantial economic and military resources. Paradoxically, the strained relations between the centre and the states is not entirely a sign of weakness. By emphasising the necessity of political participation and consensus among different regions, the fractious federal structure has successfully defused secessionist movements. While the recent unrest in the Punjab is a severe test for India's continued integrity, it is not a problem incapable of political accommodation. Moreover, India is easily the strongest power within the sub-continent, and any direct attempt at dismemberment by other powers can be virtually ruled out.

Pakistan faces more serious challenges than India both internally and externally. Even after the secession of Bangladesh, its ethnic dissensions are at least as troublesome as India's, and the largely authoritarian system of government has failed to develop roots comparable with those of India's democratic institutions. The Pakistani state machinery is strong in that it commands sufficient resources – economic, military, and political – to maintain itself against internal challenges. But its political consensus and institutions are weak, and consequently the state maintains itself at the cost of chronically unpopular and unstable governments. While credible external threats of dismemberment are easier to discern for Pakistan than they are for India, Islamabad has strong allies who are keenly interested in maintaining its existence. For this reason, any attempt at dismemberment runs a substantial risk of escalating into a wider conflict, and the deterrence value of that risk serves to offset Pakistan's greater vulnerability. Moreover, it can be argued that with the Soviets in Afghanistan, India has an interest in preserving the unity of Pakistan so as to reduce opportunities for further Soviet penetration into the sub-continent.

Both India and Pakistan can be characterised as durable state structures trying to contain fractious and not wholly stable polities. As argued during the discussion of Indo–Pakistan relations, this combination, and its intersection with the Kashmir issue, goes a long way towards explaining why the rivalry between them is also a stable feature of the South Asian complex. At least until the 1971 war,

Kashmir was the focus, though by no means the entire cause, of the Indo–Pakistani rivalry. Pakistan's repeated failure to resolve the Kashmir issue by force has, it would appear, modified its stance. Pakistan seems to have acquiesced, albeit very reluctantly, in the *de facto* status quo in Kashmir. But it must, for reasons of domestic political necessity, maintain its support for the *de jure* right of the Kashmiris to self-determination.

Following the Simla Agreement of 1972, Pakistan has gradually integrated parts of Azad Kashmir, including Gilgit Agency and Baltistan, into its own territory. Pakistan is also prepared to accept India's suggestion to convert the cease-fire line into line of actual control, with the intention that it will eventually become the recognised international boundary. However, this apparent de-escalation of tensions in Kashmir is merely a recognition of the local facts on the ground, and will not end the Indo–Pakistani rivalry. The roots of that rivalry are deeply embedded in the domestic structure and history of India and Pakistan. The very basis and principle of Pakistan's existence as a homeland for the Muslims of the sub-continent, notwithstanding the secession of the Bengali Muslims, is inimical to India's secular ideology. More crucially, Pakistan is never likely to accept a subordinate position of acquiescence to India's hegemony in South Asia. Pakistan remains committed to maintaining a balance of power in the sub-continent, even if this policy requires both a permanent arms race, and the drawing in of assistance, much to the chagrin of India, from outside powers. Pakistan's commitment to acquire nuclear options is evidence of this determination not to be overawed by India's superior resources,[1] as is the fact that nearly nine-tenths of its military forces continue to be deployed against India despite the turbulence on Pakistan's western frontiers arising from the Soviet occupation of Afghanistan.

Similarly, and in part because of the Indo–Pakistan rivalry, the position of most of the smaller states within the South Asian complex also looks stable. All but one of them are playing Pakistan's game of counterbalancing India's natural dominance by either seeking, or threatening to seek, outside supporters. Since India has neither the overwhelming degree of relative strength nor the domestic incentive to break this pattern easily, and since there is no shortage of outside powers willing to play a role, this stalemate between India and the smaller states looks set to continue. Afghanistan is the principal exception to this rule, because the consequences of the Soviet occupation cannot yet be estimated with much confidence. So far, the

Soviet move has had only a marginal impact on the essential structure of the South Asian complex, and the status quo case rests on the continuance of this trend. We will explore the alternatives under some of the other models.

These considerable local stabilities are complemented by some strong continuities in the relationship of outside powers to the South Asian region. As pointed out earlier, the United States, the Soviet Union, and China all have durable interests in South Asia which stem more from their concerns about each other than from any intrinsic interest in South Asia. None of the three great powers has a direct interest in sub-continental issues for their own sake, but for all of them South Asia is a key strategic area in the higher level game of global rivalry. The Soviets may have a locally derived interest in the form of Russia's traditional search for warm water ports, but they are obliged to be engaged in the sub-continent both to counter America's containment policy, and to promote their own containment policy against China. China's link with Pakistan also represents a vital counter-containment strategy. American ties to Pakistan derive not only from the general principle of containment, but also from the specific local interest it has in protecting access to Gulf oil.

There seems little reason to doubt that the two pairs of higher level rivalries which brought these three powers into South Asia will endure. This is especially so since China and the Soviet Union are now bound into the region not only by their rivalry with each other, but also by the Soviet entanglement in Afghanistan and by the large role that Sino–Indian rivalry plays in India's security perceptions. Both the Sino–Soviet and the Soviet–American rivalries may, as the latter has done, fluctuate between periods of confrontation and periods of *détente*. But an actual change of sides amongst them, such as occurred with China and the Soviet Union during the early 1960s, seems highly unlikely. So long as the essential rivalries continue, whether the style is *détente* or confrontation will make little difference to the durable interest of the three external powers in trying to offset the influence of their opponents in the sub-continent. Peaceful coexistence is a form of rivalry, not an end to it, and is therefore unlikely to affect the kind of intervention and manoeuvring that have characterised great power relations in South Asia.

Because the central concerns of the outside powers are more with each other than with the sub-continental states, it could even be argued that the general fact of their involvement in South Asia is more durable than the particular form of alignments that currently expresses it. There are no ideological underpinnings of significance in any of the alliances between the South Asian states and the outside powers, and such local issues as exist, like the Sino–Indian border dispute, are not beyond resolution. To that extent, they are all alliances of convenience reflecting classical balance of power behaviour. Following that classical logic, and keeping in mind the various interests of both the local and the external states, it is not all that difficult to imagine a scenario in which the Soviet Union is aligned with Pakistan, and China and the United States are aligned with India.

Historical momentum and memory may stand as a barrier to any such shift, but it can be conjectured convincingly that in many respects, such a set of alliances might serve the interests of all the parties as well as, or even better than, the current one. A Soviet–Pakistan alignment would take pressure off both sides from the situation in Afghanistan. It would give the Soviets a friendly corridor to the Indian Ocean, and it would get Pakistan out of the squeeze it is now in between the Soviet Union and India. A Sino–Indian alignment would strengthen China's position against the Soviet Union, and release India from the fear of two-front wars. On the face of it, such an alternative set of partnerships might even be more in harmony with the domestic political structures of the states concerned than the existing arrangement. It is a commonplace observation that the United States and India are the world's two largest democracies, while the Soviet Union and Pakistan share not only the form of a repressive state structure legitimised by an official ideology (albeit not the same ideology), but also a political system dominated by its largest ethnic group.

From the perspective of our models, however, even a transformation of external links on as large a scale as this would not disturb the continuity of the status quo within South Asia, because it would not change the essential structure of relations within the local security complex. Nothing could illustrate better than this speculation the impact and the importance of the local and regional patterns of security relations in the larger picture of international relations as a whole. The local structures are durable, and they play a crucial mediating role in relations between the great powers and the local states.

The status quo can also incorporate some shifts in areas where change is either currently underway or else looks likely. Up to a point, for example, the continued development of links between

Pakistan and the Arab states, particularly Saudi Arabia, need not challenge the status quo. So long as each side simply draws material support from the other, and does not engage directly in either its patterns of hostility or its conflicts, there will be no compelling reason to dissolve the boundary between the Gulf and South Asian complexes. Similarly, the continuity of the status quo in South Asia would not necessarily be disturbed either by the continuation of moves to improve Sino–Indian relations, or by the coming to fruition of Bangladeshi and Sri Lankan moves to cultivate closer ties with the United States.

INTERNAL TRANSFORMATION

None of the most obvious options for South Asia suggested by this model seem likely to occur. It has been argued that neither India nor Pakistan is anywhere near the brink of break-up as a result of internal secessionist movements, even though Pakistan's unity is under considerable strain in Baluchistan and the Northwestern Frontier. Despite optimism among some South Asian analysts, the break-up of Pakistan in 1971 did not make much impact on the distribution of power in the South Asian complex, and did not leave India in a position of undisputed regional dominance. Even the occurrence of highly unlikely events, such as a successful secession by one or more of India's troubled provinces, or the establishment by the Sikhs of an independent Khalistan, would likewise fail to transform the power structure of the complex.

For the reasons argued above, a change in the complex-defining pattern of hostility between India and Pakistan is also hard to imagine. Perhaps the only foreseeable event capable of reconciling the two local powers would be a massive unilateral attempt by the Soviet Union to extend the range of its direct occupation in South Asia. Such a move is unlikely, *inter alia*, precisely because it would put at risk the alliance with New Delhi to which the Soviets have devoted so much time, and so many resources, to cultivating over the last quarter century. While the Indians have gained much from the alliance with the Soviet Union, the relationship is not one-sided. The Soviets have few reliable Asian allies, and Moscow can ill afford to exchange the cordiality with New Delhi for the doubtful advantage of occupying further South Asian territory.

In theory, the most likely path to internal transformation in South

Asia has always been a steady shift in the distribution of power away from bipolarity towards a unipolarity centred on India. Many Indians have hoped, and many Pakistanis dreaded, that such a shift would occur naturally. But no such transformation has taken place, and none is in the offing. India's non-absorption of the smaller states is explained both by a lack of enthusiasm in New Delhi for the consequences of doing so (even assuming that the absorption itself would be easy), and by fear that any direct attempt at expansion would draw outside actors into the sub-continent (and therefore would not be easy). Bangladesh is a liability whose possession India cannot logically relish: New Delhi wants no more Muslims and Bengalis. Nepal and Bhutan derive considerable leverage from the Sino-Indian rivalry, and this enables them to keep any threat of direct Indian dominance at bay. India's stand-off against Pakistan is explained by Islamabad's success in bolstering its power resources sufficiently to maintain its status as India's principal rival, a success likely to be enhanced by the looming nuclear factor in relations between the two countries. All of these situations appear durable.

The only exceptions to India's reluctance to exploit its power by dominating or absorbing its smaller neighbours have been its annexations of Sikkim and the small Portuguese colonies, its dominant relationship with Bhutan (though here the trend is in the opposite direction), and, arguably, its position in Kashmir. For the rest, New Delhi has tolerated, and even encouraged as in the case of Bangladesh, the existence of small, independent states on its borders, even when these have strong ties into its own domestic politics because of overlapping ethnic and religious groups. On the face of it, such tolerance sits oddly with the combination of great disparity in power, a widespread sense within India of an intimate sphere of influence embracing the whole of the sub-continent, a long history of regional unity both before and under the British Raj, and a secular federal political constitution that could, in theory, accommodate the absorption of neighbouring kindred peoples.

Three factors explain India's lack of enthusiasm for the consequences of expansion. First, overt expansionism would clash with New Delhi's long-held foreign policy positions on anti-colonialism and non-alignment. India's reputation already suffers because of the wars with Pakistan, and the government is not anxious to increase its burdens in this regard. Second, the existing pattern of relationships with the small states is not highly aversive to India, and the incentive to embark on major change is therefore low. New Delhi has considerable economic leverage, and the smaller states do not go out of their way to challenge its core interests.

Third, and in some ways most important, is an understandable reluctance to loosen and dilute further India's already fractious polyglot federation. As seen from New Delhi, some of India's domestic politics must look little different from foreign affairs, in the sense that they involve the option of using force against actors who treat themselves as sovereign and who are prepared to use force to defend their claims. From such a perspective, the incorporation of further problematic populations into the Indian body politic may well look less appealing than dealing with them as weak neighbours. The force of this logic is illustrated in reverse by the contrast between the current trend towards quite good and unproblematic relations between Bangladesh and Pakistan, and their catastrophic relations when they were bound together within a single state. The absorption of the smaller states by India would make only a marginal difference to the power structure of the South Asian complex, but it would almost certainly add enormously to the problems of internal governance. It would also increase tensions between India and Pakistan by signalling an imperial mentality in New Delhi towards the continued existence of Pakistan.

In addition to these disincentives, India is constrained by the fear that any attempt on its part to extend control over the smaller states would be met by them calling in external powers to redress the balance. Any increase in penetration of the subcontinent by external powers almost automatically works against New Delhi's own drive for regional and global status. Once the Indian government acquiesced in China's occupation of Tibet, it sealed the elimination of the traditional buffer that would have allowed it more leverage over Nepal and Bhutan. Both of these states have played the China card effectively in their relations with India. Bangladesh and Sri Lanka have also demonstrated their willingness to cultivate great power ties as an offset to India's influence.

Pakistan's success in offsetting India's larger resources has rested primarily on its ability to attract substantial support from outside powers. There are three reasons for thinking that this state of affairs will continue. First, some of Pakistan's current supporters are committed to it for durable reasons of their own. The United States may cool to Islamabad as it has done periodically in the past, but Saudi Arabia's backing rests firmly on Islamic ties and calculations of mutual interest in domestic and military support. China's support is guaranteed so long as the Indo–Soviet alignment remains in force. Second, as argued in the previous section, the United States, the Soviet Union and China are all committed to South Asia for reasons arising from their own rivalries. Because of this situation, Pakistan will be able to find a great power backer virtually regardless of what shifts in alignment occur. As outlined above, it is quite possible to envisage a Pakistani alliance with the Soviet Union, the absence of any ideological or internal obstacles to such a move being demonstrated by the flirtation of the two during the late 1960s. The third reason relates to this second one inasmuch as it concerns Pakistan's highly pragmatic attitude towards alliance making. Given Islamabad's versatility, and despite its ostensibly ideological character as a state, there are few countries with which Pakistan would not ally itself if there was advantage to be gained. By giving primacy to maintaining its status against India, Pakistan has put itself in a flexible position to exploit any outside support it can find. The period during the 1960s when its two major allies, China and the United States, were still treating each other as mortal enemies is instructive in this regard.

The looming nuclear parity between India and Pakistan, whether at threshold or weapon status, also works against Indian hegemony. At this local level, even a few nuclear weapons would introduce significant deterrent effects, and the uncertainties of threshold status would force both sides to assume some nuclear risk in any future war between them.

For all these reasons, an imbalance of power is unlikely to be the mechanism of an internal transformation of the South Asian complex. So long as the Pakistan state can prevent its own demise from internal pressures, the only feasible path to internal transformation in South Asia lies through the unlikely route of a joint assault on Pakistan by India and the Soviet Union. It is possible to find motives for such an action, and it serves Pakistan's interests in maintaining American support to draw attention to them. Many in India would not be averse to eliminating Pakistan as a rival, since this would give India much greater freedom of manoeuvre on both the regional and the international stages. The Soviet Union could certainly see advantage in having a corridor of compliant allies stretching from the Hindu Kush to the Arabian Sea.

However, a direct military invasion to this end, even if jointly undertaken, would be unappealing on grounds both of the risk of escalation, and the political unacceptability of the action to domestic and international audiences. Both China and the United States would be under intense pressure to prevent the dismemberment of Pakistan, and the whole scenario has the makings of a global crisis on a scale not experienced since 1941. The likely risks are wholly disproportionate to the prospective gains. A more subtle encouragement of separatism among Pakistan's Baluch, Pathan and Sindi minorities might bear fruit, especially if the dominant Punjabi core of Pakistan becomes deeply divided on the question of the political constitution of the state. Pakistan might not be able to resist a sustained political assault, and its allies would be hard put to find ways other than military, of helping it. Both India and the Soviet Union might find useful pickings in the balkanisation resulting from an internally generated second partition. The idea of a Greater Afghanistan could be revived to legitimise some annexations, and either an expanded Afghanistan under internal rule, or a cluster of small successor states might serve as acceptable buffers between India and the Soviet Union.

This scenario, while not impossible to conceive, would require a level of aggressive expansionism on the part of both the Soviet Union and India which would be a very substantial departure from their current norms of behaviour. It would also require a major deepening of their alliance, a willingness to intensify greatly the hostility of China, and the occurrence of favourable conditions in Pakistan. Even if the Soviet Union really wants to extend its control in South Asia, the costs and requirements of this scenario would still render it unappealing. Moscow could more reasonably pursue such goals by attempting to cultivate a close relationship with Pakistan, even at the expense of its alliance with India. For India, the scenario raises other serious risks. Could New Delhi prevent secessionist fever spreading from a dismembered Pakistan into its own vulnerable body politic? Would the gains to India offset the loss of part of its sphere of influence to the Soviet Union, and does India's alliance with the Soviet Union mean that it would welcome the long-term prospect of living with a lengthy shared border? How would India deal with the increased ties between the smaller states and other outside powers which would be the almost inevitable response of the smaller states to such a move?

For all these reasons, this scenario for internal transformation looks as unlikely as all of the others we have surveyed. It would be a medium term option at the earliest, and should it occur would raise issues for the next model, external transformation.

EXTERNAL TRANSFORMATION

Three issues need to be considered under this heading: the status of Afghanistan, the significance of Pakistan's links with the Gulf, and the prospect of the South Asian complex being subordinated to a Sino–Indian one.

The issue of most immediate interest is Afghanistan. In terms of our four models, the Soviet occupation of Afghanistan can be seen either as an overlay of part of the South Asian complex, or else as an external transformation of it (i.e. a change in its outer boundary) resulting from the effective annexation of one of its peripheral members by an outside power. Both interpretations contain elements of truth, and both partially distort the reality. Since the situation is still in flux, and Soviet intentions for the long term are not yet clear, it would be fruitless to try to judge in favour of either interpretation.

The merit of the partial overlay view is that it leaves Afghanistan still within the South Asian complex, thus allowing an element of foreign domination to be combined with continued definition of the Afghan–Pakistan issues as local. The problem with it is that the overlay model was not designed to be applied only to a marginal part of a complex, and so gives no clear image of the impact on the complex as a whole. The logic of the overlay model was designed principally to cover interventions on a much larger scale.

The merit of the external transformation view is that it captures the sense of a major southward extension of the Soviet Union's security periphery. In some important respects the Afghan borders with Iran, Pakistan and China are now Soviet borders, and to that extent, Afghanistan is no longer a local state in the normally accepted sense. What were previously local issues between neighbouring states have acquired a new level of significance because of their absorption into the direct strategic concerns of a superpower. Given Pakistan's alignment with China and the United States, the major rivals of the Soviet Union, and its rivalry with the Soviet Union's major independent ally, India, the external transformation view is particularly salient for Islamabad. The danger in this view is that it exaggerates the extent to which Afghanistan has been cut off from South Asia, and therefore obscures the reality of continuing local level security interactions.

As things currently stand on the Afghanistan question, the models are more useful in helping us to ask questions about the future than they are in enabling us to define present conditions. If the Soviets stay in direct military occupation of Afghanistan, then the situation will evolve into an external transformation. If they withdraw, leaving behind a relationship modelled on the Warsaw Pact, then partial overlay would be the best model. If the Soviets try to extend their direct physical control further into South Asia, then massive changes within the terms of all four models become a real probability.

The situation in Afghanistan bears on, but is by no means the only factor in, the question of external transformation in the form of a merger of the South Asian and Gulf security complexes. As argued under status quo above, Pakistan's links with the Gulf states rest on a number of apparently durable local interests: religious, economic, military and political. For Pakistan, the link is a welcome addition to its array of external supporters, and one with useful parallel connections to the United States. Given the importance of the Islamic idea to Pakistan's political constitution, it is hard to imagine circumstances in which Islamabad's ties to the Gulf would be dropped or substantially downgraded, though entanglement in a Gulf conflict might force it to take a less even-handed role than it now does. The important questions are, therefore, about the significance of the current level of linkage, and the circumstances that would be likely to increase it.

The interesting thing about the current ties between Pakistan and Saudi Arabia is that they have so far been oriented primarily towards the general domestic and military strengthening of the two states without either of them becoming directly engaged in the other's pattern of hostilities. Pakistan is not a member of the Gulf Cooperation Council, is not deeply engaged in inter-Arab politics, and has not been much engaged in the Iran–Iraq war despite sharing a substantial border with one of the belligerents. Saudi Arabia has not taken on any hostility towards India, though it has sided with Pakistan in opposition to the Soviet occupation of Afghanistan. Thus although there is some significant linkage between the two local complexes in terms of shared sources of power between some of their members, these links do not yet include any significant merger of security perceptions in terms of common rivalries or hostilities. So long as this remains the case, there are no real grounds for arguing that the boundary between the two local complexes has dissolved. The truth of this interpretation can only be tested if either Pakistan or Saudi Arabia get involved directly in a war. Only if their relationship did not remain at arms length, or if they concluded a formal alliance, would the inter-complex boundary be brought seriously into question.

In the longer run, a number of external developments could push Pakistan towards closer involvement with the Gulf states. American policy has linked Pakistan to the Gulf since the early 1950s in the context of a declared 'region' labelled Southwest Asia, and through the (now defunct) machinery of the Central Treaty Organisation (CENTO). This pressure has doubtless contributed to the current ties, but paradoxically, it could well be a lowering of American interest in the area that would do most to encourage stronger links. Pakistan's prime need is to offset India, and any lessening of support from either of its major external backers would create pressure on Islamabad to strengthen its remaining ties. Thus either a Sino–Indian *rapprochement* or a waning of American commitment to the Indian Ocean could force Pakistan to turn more to the Gulf. Increased Soviet pressure from Afghanistan could also make Pakistan more dependent on all of its backers.

The same effect might result from the quite different dynamics of Pakistan's nuclear program, and its connections to some Arab states. If Pakistan finds itself engaged in a full scale nuclear rivalry with India, it could easily become more and more financially dependent on, and therefore politically obliged to, its Arab backers. Such a development could manifest itself in two ways: it could increase the pressure on Pakistan to transfer nuclear weapons or technology to Arab states, or nuclear status could enable Pakistan to play a more active role in the security affairs of the Gulf. Even at the current levels of linkage, Pakistan risks finding that the convenient fund-raising slogan of an 'Islamic bomb' has brought its nuclear program directly into the realm of Israel's security perceptions.[2] Any dramatic increase in Arab access to Pakistan's nuclear capability by either of the above routes would certainly engage Israel's attention. The potential thus exists here for some quite dramatic fusions of South Asian and Middle Eastern security interests as a result of developments in the nuclear arms race between India and Pakistan. In terms of both threats and resources, the nuclear issue could lead to strengthening ties between the South Asian and Middle Eastern security dynamics.

The issue of a merger between the South Asian and Middle Eastern security complexes therefore remains an open question for the future. Superpower rivalry over Afghanistan certainly encourages the development of ties between Pakistan and the Gulf. But short of major Soviet attempts at further southward military expansion, superpower influence alone seems unlikely to break down the existing structure of local security complexes. Should such a transformation come about, it is much more likely to result from the spreading security consequences of Pakistan's increasingly difficult struggle to maintain its status against the greater weight of India.

Also worth noting is the remote possibility of the rivalry between India and China rising to such a pitch as to redefine the security complex on a much larger scale. Current trends do not point in this direction, with both India and China pursuing separate interests, and seeming much more interested in establishing cordial relations than in pursuing rivalry. Indeed *détente*, or even *rapprochement*, seem more likely options than major rivalry, though bets on any trend are rendered hazardous by the uncertainties of Pakistan, which could easily inflame Sino–Indian relations because of their opposed commitments there. Deficiencies in the power capabilities of both countries make a major regional rivalry between them unrealistic at this time, as does the fact that both have more pressing security problems in other directions. Such a transformation could only occur if both countries were absolutely and relatively much more powerful than they are now. Both would also need to be much more externally directed than they are at present. Some of the potential here is illustrated by India's political support for Vietnam, which could be an early pointer to a future spheres of influence rivalry between India and China in South-east Asia. Given the new Soviet position in Afghanistan, joint Indo–Soviet pressure on Pakistan could eventually resolve the South Asian complex in India's favour, leaving it secure enough to attempt a broader extension of its regional influence. These events are for the next century if at all, but they illustrate one possibility for a major external transformation of the South Asian complex several decades hence.

OVERLAY

Despite the consolidation of the partial Soviet overlay in Afghanistan, and increases in both American and Soviet military presence in the Arabian Sea area since the late 1970s, it cannot be said that overlay is a very realistic option for South Asia. Only the two superpowers are capable of exercising the overlay option, and any thoughts they might have in that direction are focused much more on the Gulf than on South Asia. Pakistan falls within their area of interest, and the vulnerabilities of its domestic political structure make it a feasible target. India is clearly not within the area of intense superpower concern, and is too big and too powerful to be a target for large scale, long-term direct intervention.

Within the area of the Gulf/South-west Asia, none of the local states would be receptive to direct foreign presence on the requisite scale, and the virulence of the local disputes would be difficult to suppress. In addition, existing patterns of alignment between local and external states do not produce clean lines of division like those that were available to the superpowers in Europe and North-east Asia. To make the necessary changes would require a level of intervention so high as to raise the unacceptable risk of direct clashes between the superpowers. Partial overlay options involving Pakistan and Iran are conceivable in the context of a Soviet drive for a corridor to the Arabian Sea. But such a drive implies a Soviet willingness to risk direct clashes between Soviet and American forces. This kind of risk-taking is unprecedented in the post-war world, and the benefits to the Soviets from taking such a gamble do not compare favourably with the risks of nuclear war, and the alienation of world opinion, both of which such action would entail.

The Soviet Union is additionally constrained by the fact that any further unilateral Soviet move into the region would intrude into India's sphere of influence, and so place at risk the Indo–Soviet alliance. The Soviets' need to maintain the alliance with India makes their occupation of Afghanistan look more like an end in itself than like the beginning of a new imperial expansionism.[3] The overlay scenario explored above in which India and the Soviet Union collaborate in a carve-up of Pakistan is unlikely for the reasons already argued. But should New Delhi and Moscow be seriously considering such action, they would have to take account of a closing 'window of opportunity' relating to Pakistan's nuclear status. Once Pakistan is known to possess nuclear weapons, any project involving its dismemberment becomes much more dangerous to contemplate.

CONCLUDING OBSERVATIONS: INSECURITY AND ALIGNMENT

The framework of the study makes it obvious that the security of the South Asian states is heavily bound into a complicated and far-reaching international context over which the local states have limited control. But although the local states cannot control the security environment created by the great powers, it is important to note that the independent pattern of local security relations plays a major role in shaping the character of external penetration into the region.

The dynamics that produce alignment with external powers result both from the pull of the local states and the push of the great powers. The local states are usually motivated by considerations arising from local rivalries, while the great powers are mainly motivated by their rivalries with each other. The great powers use the local cleavages to facilitate their entry into a region even though they are not particularly interested in the local rivalries themselves. The mismatch of interests which results can make the local–external alliance links unstable, as illustrated elsewhere by cases like Egypt, Ethiopia, and Somalia. As speculated above, South Asia might also be vulnerable to such external alliance switching. In the South Asian case, however, the Indo–Soviet alliance rests on two linked mutual interests: both countries have security concerns about China, and both have an interest in developing India as a strong regional power. This has made their alliance more stable than it would otherwise have been, but there is enough room for movement in both India's relations with China, and the spheres of influence conflict between India and the Soviet Union over Afghanistan, to raise doubts about the continuity of that stability in the future.

The internal structure of the South Asian complex is more durable than the pattern of external alliances that currently reinforces it. Even massive shifts in the pattern of external alliances would not change the essential structure of the South Asian complex. Indeed, it is the very durability of the local structure that virtually guarantees the continuance of *some* major pattern of alignment with external powers, even if not the existing one. Alignment, as we have seen, reinforces the essential structure of the subordinate complex by raising the power resources available to the local states, but without controlling or mitigating the local patterns of rivalry and hostility. Since the external powers depend on those local patterns to gain access, they have both limited incentives and limited ability to moderate them. The failure of the Soviet Union during the middle and late 1960s to cultivate successfully close relations with both India and Pakistan is instructive in this regard.

It is difficult to argue decisively that great power involvement in

South Asia has either increased or decreased security. It has done both. The smaller states have been the clearest beneficiaries in that external alignment, or the threat that they might resort to it, has enabled them to enjoy greater independence than would otherwise have been likely. Pakistan has also been a main beneficiary, though it can hardly be described as secure. Without the major inputs of external resources that it has received, it is doubtful that Pakistan would have been able to maintain its rivalry with India nearly so successfully as it has. But the term 'success' has to be used here with caution. Pakistan still exists, and it still avoids subordination to India. But its pretensions to being the Islamic homeland for the subcontinent lie in ruins as a result of the Bangladesh secession, and it has failed to oust the Indians from Kashmir. In addition, it can be argued that Pakistan has paid a high political and social price domestically for the external support it has received.[4]

Here the argument comes full circle, for by supporting Pakistan for their own purposes, the United States and China have made a major contribution to sustaining the bipolar power structure in South Asia. Without that structure, their own game in the region would be much more difficult to play. External alignment helps to maintain the structure of the local complex, and the continuity of that structure ensures that the region remains continuously open to external rivalries.

Alignment thus only seems to relieve the problem of insecurity. In terms of individual states, alignment may be vital to their ability to maintain a certain definition of national security. But in terms of the regional subsystem as a whole, the case of South Asia suggests that alignment works powerfully to sustain the very conditions that define the insecurity of the local states. It does so by cutting off the two main paths to regional stability. By making relatively large financial and military resources available to local states, the mechanism of alignment both prevents the resolution of local disputes, and blocks the emergence of locally dominant regional powers.

Because the local states are key players in the alignment game, we can conclude that they do have considerable *influence* over how external powers impinge on their affairs. But they have little ability to *control* external penetration into the region because they are unable to resolve the local rivalries which generate the demand for external support. Conversely, the great powers cannot either influence or control the local security dynamic because they depend on it for access to the region. The process of rivalry amongst the great powers automatically reinforces the existing local security structure, in a pattern that is extremely difficult to break. Any attempt to stem external penetration by forming a local security community has to overcome not only the local rivalries but also the global ones. What can break the cycle is if one or more of the great powers decides to expand directly into the local complex. Overlay, as the case of Europe demonstrates, is a possible third path to regional stability. It is for this reason that the Soviet occupation of Afghanistan is so important to the future security of the South Asian states.

Finally, one cannot help but notice the extraordinary centrality of Pakistan in this study. On almost every issue, and within almost every interest, Pakistan lies at the heart of the possibilities for change. It is the least internally stable of the major states in South Asia. It maintains, against the odds, the structure of the South Asian complex. With the Soviet move into Afghanistan, it lies at the focal point of rivalry among the three external great powers active in the region. It provides the Islamic linkage between South Asia and the Middle East. And it is the nuclear aspirant most likely to entangle the affairs of South Asia with those of the Arab world. The future pattern of security in South Asia very much rests on the policy and on the fate of the widely unloved regime in Islamabad.

NOTES

1. On this topic, see Bhabani Sen Gupta, *Nuclear Weapons? Policy Options for India* (New Delhi, 1983) chs 1–3.
2. Ibid., ch. 3.
3. The delicacy of Indo–Soviet relations on the issue of Soviet penetration into South and Southwest Asia is extensively discussed in Timothy George, Robert Litwak and Shahram Chubin, *Security in Southern Asia 2: India and the Great Powers* (Aldershot, 1984) pp. 71–145.
4. See, for example, J. Ansari and M. Kaldor, 'The Bangladesh Crisis of 1971', in Mary Kaldor and A. Eide (eds), *The World Military Order* (London, 1979) pp. 136–56.

Index

Abdullah, Sheik, 100, 109, 118
Aden, *see* People's Democratic Republic of Yemen
Afghanistan, 3, 8, 10, 14, 16–17, 19, 25, 29, 31, 62, 72–4, 78, 80–1, 83, 119–20, 122, 128–30, 146, 148–53, 170, 173, 177, 181, 183, 187, 198, 200–1, 212, 217, 220–3, 225–8, 236–9, 244–9, 252
Africa, 12, 18, 19, 29, 84
Akali Dal, 51–3
Ali, Choudhury Rahmat, 66
Ali, Mohammed, 184
All Assam Gana Sangram Parishad (AAGSP), 49
All Assam Student's Union (AASU), 49
Anandpur resolution, 51–2
Andhra-Pradesh, 42, 43, 46, 57
Arab–Israel complex, 28
Arab–Israel war (1967), 171
Arab League, 172
Arunachal Pradesh, 48
ASEAN, 9, 226
Asian supercomplex, 16, 18, 158
Assam, 38–9, 44, 46, 47, 48–51, 56, 130
Austria–Hungary, 22
Awami League, 85, 114–15, 132

Baghdad Pact, *see* CENTO
Bahrein, 10, 167, 170
Bakshi, Ghulan, 110
Balance of power, 7, 12, 28, 122
Baluchis, 61, 71, 73–4, 84–5, 173, 244
Baluchistan, 66, 72, 74, 76–8, 189, 197, 212, 240
Bandaranaike, S.W.R.D., 140–1

Bangladesh, 9, 10, 16, 22–3, 48, 49, 65, 69, 71, 77, 79, 83, 85, 95, 114, 118–21, 127–36, 139, 141–2, 145, 151, 153, 157, 169, 181, 195, 197, 226, 236, 241–2, 251
Bengal, 66, 95, 99, 102, 143
Bengalis, 76, 81, 84–5, 114–16, 129–34, 192, 241
Bharatiya Janata party (BJP), 44
Bhartiya Lok Dal (BLD), 46
Bhindranwale, Sant, 52–3
Bhutan, 10, 49, 127, 139, 142, 145–8, 152, 157, 241–2
Bhutto, Mumtaz Ali, 77
Bhutto, Zulfiqar Ali, 69, 77–8, 81, 87, 101, 105, 107–9, 111–13, 116–18, 121, 188, 191, 195
Bihar, 42, 130, 143, 145
Brecher, M., 6
Brezhnev, L., 197, 225
Britain, 24, 93, 97, 100, 104–5, 107, 130, 138, 140–2, 148, 153, 163–4, 171, 189, 193, 200, 208, 209–11
Buddhists, 136
Burma, 9, 10, 14, 15, 48, 49, 128, 136, 146

Cantori, L., 6
Caste, 38, 41–2, 94–5
Central Treaty Organisation (CENTO), 165, 169, 182, 193, 208, 212, 247
Chad, 14
China, 8, 11, 14, 15, 16, 17, 24, 49, 62, 83, 104–6, 108, 115–16, 120, 122, 130, 136, 139, 141–4, 146–7, 149–53, 157–8, 174, 181–201, 205, 208–23, 225–7, 238–9, 242–5, 248, 250–1

Civilisational area, 10
Colombo Plan, 147
Commonwealth, 140, 213
Congress party, 37, 40, 41, 45, 51, 66–7, 110
Congress (I) party, 42–4, 52, 56–7
Czechoslovakia, 186, 225

Desam, Telugu, 43, 57
Diego Garcia, 139, 221
Disturbed Areas Act, 48
Dravida Munetra Kazhan (DMK), 38
Durand line, 72

East Bengal, 67–8, 76–7
Economic security, 13
Egypt, 25, 26, 29, 100, 171, 210, 221, 250
English, 39
Enlai, Zhou, 184–5, 192
Essential structure, 22–3, 27–9
Ethiopia, 11, 26, 221, 250
External transformation, 27, 245–8
Europe, 10, 13, 18, 24, 28, 30, 79, 249, 252
European security complex, 26, 28

Farakka Barrage, 134–5, 214
France, 14, 25, 171, 193

Gandhi, Indira, 43, 46, 54, 116–17, 119, 150, 194, 197, 218
Gandhi, Mahatma, 63
Gandhi, Rajiv, 53–4, 56, 131, 140, 150
Germany, 22, 25, 28, 30, 200
Goa, 22, 141, 157
Government of India Act (1935), 68
Greece, 26, 209–10
Gujarat, 42, 43
Gulf, 13, 16, 21, 23, 82, 120, 148, 158–9, 161–8, 173–4, 212, 249
Gulf Cooperation Council (GCC), 164–5, 167–8, 176, 246
Gulf security complex, 9, 10, 11, 14, 15, 161–68, 176, 240, 246
Gurkhas, 130

Haas, M., 6
Haq, Mahbub ul, 82
Haq, Zia-ul, 70–1, 79, 81, 86–7, 118, 172, 217
Harijans, 42
Haryana, 38, 42, 43, 52
Hasan, Syed Zafarul, 66
Heng Samrin, 15, 198
Himachal Pradesh, 43, 52
Hindi, 38–9
Hindus, 39–40, 53, 63–5, 93–4, 97–8, 132, 136, 143
Hormuz, Straits of, 164, 168
Horn security complex, 11
Husain, Chulan, 68
Hyderabad, 98

India, 8, 9, 10, 11, 13, 15, 16, 22–3, 37–57, 65, 77, 80, 83, 91–2, 98, 102–3, 128–33, 136–53, 157, 169, 173–6, 181–5, 187–201, 207–28, 235–50; economy, 54–6, 113, 174–5, 224–5
Indian Army, 53
Indian National Congress, 84, 93–4, 96–7, 208
Indian Ocean, 220–21, 228
Indonesia, 100, 107
Indo–Nepalese Treaty of Peace and Friendship (1960), 143–4, 146
Indo–Pakistan rivalry, 9, 14–15, 19–20, 25, 27–8, 61–2, 83, 91–123, 127–30, 150, 152, 182–3, 187, 196–9, 202, 208, 213–21, 236–7, 240–4, 249–51; nuclear weapons, 119–21, 150, 175–8, 199, 201, 217–18, 222, 237, 243, 247, 249
Indo–Pakistan war (1947), 99
Indo–Pakistan war (1965), 20, 111–13, 120, 170, 187, 190, 195, 211
Indo–Pakistan war (1971), 20, 114–18, 120, 190–2, 194–6, 212, 225, 236
Indo–Soviet relations, 18, 19–20, 183, 187–9, 192–6, 198–200, 213–21, 224–7, 240, 244, 249

Index

Indo–Soviet alliance, 20, 115, 191, 194, 215, 225, 244, 249, 250
Indo–US relations, 211–12, 215, 220–23
Internal transformation, 27–9, 240–4
International Atomic Energy Agency (IAEA), 223
International Monetary Fund (IMF), 55
International system, 5–6
Iqbal, 63–4, 66
Iran, 3, 8, 9, 11, 16, 23, 25, 26, 29, 62, 74–5, 77, 107, 120, 159, 164–7, 169–70, 173–5, 187, 209–10, 222, 226–7, 245, 249
Iran–Iraq rivalry, 74, 171
Iran–Iraq war, 164–5, 167, 172–3, 224, 246
Iraq, 9, 25, 29, 164–8, 173, 210, 218
Islam, 63–5, 71, 75, 84, 94–5, 131–2, 135, 149, 159–60, 169, 171–2, 176, 182, 205
Israel, 11, 25, 26, 28, 29, 138, 161, 165, 169, 171, 175, 178, 210, 218, 247
Italy, 22, 28

Jaamat-E-Islami, 40
Jackson, Robert, 114
Jaminat-i-Ulema Islam (JUI), 78
Jammu, 41, 46, 47, 57, 97, 100–1, 105, 182
Jana Sangh, 46
Janata party, 43, 46
Japan, 14, 18, 23, 30, 171
Jayewardene, Junius, 138
Jinnah, Mohammed Ali, 63, 65, 67, 96, 99
Jodhpur, 98
Junagadh, 98, 102

Kampuchea, 15, 146
Karnataka, 43, 46, 57, 129
Kashmir, 14, 41, 46, 47, 57, 97–114, 116–19, 122, 150, 169, 182–5, 190, 193, 195, 197, 210, 213–14, 220, 236–7, 241, 251
Kenya, 221
Kerala, 42

Khaliquzzaman, Chaudhuri, 67
Khan, Khan Abdul Qaiyyum, 78
Khan, Khan Abdul Wali, 78
Khan, Ayub, 68–70, 81–2, 86–7, 100–1, 103–9, 112–13, 115
Khan, Liquat Ali, 67–8, 100, 182, 209–10
Khan, Sikander Hyat, 66
Khan, Yahya, 69, 112, 115, 191–2
Khrushchev, Nikita, 183, 213
Korean war, 182, 185, 209
Kurds, 165
Kuwait, 10, 164–5, 167–8, 170–72

Laos, 14
Latif, Syed Abdul, 66
Lebanon, 14, 172
Libya, 29, 175
Lok Dal, 44

McGhee, George, 209–10
Maghreb security complex, 11
Maharashtra, 38, 43
Maldives, 10
Manipur, 48
Marxism, 65, 149
Maxwell, Neville, 107, 110, 189
Meghalaya, 48
Menon, V. P., 99
Middle East, 10, 24, 29, 121, 158, 160–1, 173–5, 209–10, 222–4
Middle Eastern complex, 11
Middle Eastern supercomplex, 11, 15, 16, 25, 28
Mirza, Iskander, 68
Mizoram, 48
Mizos, 48, 130, 135, 189, 199
Mountbatten, Lord Louis, 97–100, 102
Muhammed, Ghulam, 68
Musa, Mohammed, 107
Muslim League, 65–9, 84, 97, 208
Muslims, 39–41, 63–5, 93–8, 103, 109, 121–2, 131–2, 237, 241
Muslim Personal Law, 40
Mysore, 38

Nagaland, 48
Nagas, 48, 189, 199

National Awami party (NAP), 77–8
National Democratic Alliance, 44
Nationalism, 79–80, 84, 95
NATO, 30, 210
Nehru, Jawaharlal, 39, 45, 70, 96, 99–100, 102–4, 108, 110, 140, 182–3, 185, 189, 213
Nehru Report (1929), 96
Nepal, 10, 128, 130, 139, 141–8, 152, 157, 192, 241–2
Nonalignment, 104, 133, 182, 213, 221
Nonaligned Movement, 107, 147, 157, 190
North American complex, 28
Northeast Asia, 24, 249
Northwest Frontier Province, 66–7, 72, 76–7, 96, 99, 197, 240

Oman, 164, 166–8, 170–2, 221
Overlay, 24, 26, 27, 30–31, 245, 248–9, 252

Pakistan, 3, 7–10, 13, 14, 16, 19, 21–5, 27–8, 53, 60–87, 91–2, 101–2, 121, 128–31, 135, 139, 141, 148–53, 157–9, 169–77, 181–5, 187–201, 207–28, 236–52
Pakistan–Afghanistan dispute, 61–2, 72–4, 148–51, 209, 245
Pakistan Peoples' Party (PPP), 77–8
Pakistan–Soviet relations, 192–5, 200–1, 218–9, 244
Pakistan–US relations, 120–21, 187, 200, 208–13, 217, 221–3
Palestine Liberation Organisation (PLO), 171
Panikkar, K. M., 140
Partial Test Ban Treaty (1963), 186
Partition, 96–7
Pathans, 61, 71–3, 84, 189, 244
People's Democratic Republic of Yemen (PDRY), 166, 171–2, 221
Philippines, 171
Plains Tribal Council of Assam, 50
Poland, 62
Punjab, 38, 41, 44, 46, 47, 51–4, 56, 67–8, 95, 99, 102, 150, 216, 236
Punjabis, 61, 71–2, 74–7, 84–5, 244

Qadri, Afzal Husain, 66
Qatar, 10, 167, 170
Qureshi, Ishtiaq Husain, 65

Rahman, Sheikh Mujibur, 114, 132–5, 197
Rajamannar Committee, 45
Rajasthan, 52
Rann of Kutch, 14, 110–111, 190
Rashtriya Sangstha Sangha (RSS), 40
Reagan, Ronald, 120
Regional Cooperation for Development (RCD), 107
Regional security, 3–5, 7, 9
Regions, 6–7
Rusk, Dean, 105
Russett, B., 6

Sandys, Duncan, 105
Sarkaria Commission, 47, 52
Saudi Arabia, 9, 25, 107, 159–77, 205, 246
Saudi–Pakistan relations, 169–77, 242
Sayed, G. M., 64
Second World War, 24, 26, 30
Security complex, 8–11, 21;
higher-level, 11–21;
lower-level, or local, 11, 13, 14, 21
Seven United Liberation Army (SULA), 48
Seychelles, 221
Shastri, Lal Bahadir, 108, 111, 113
Sikhs, 39, 41, 51–4, 80, 130, 240
Sikkim, 46, 119, 127–8, 130, 142, 144, 147–8, 152, 157, 190, 198, 241
Sind, 66–8, 72, 74–7, 96, 110
Sindhis, 61, 71, 73–5, 77, 84–5, 244
Singh, Maharaja Hari, 97–100
Singh, Sardar Swaran, 105, 109
Sinhalese, 136–8
Sino–American rapprochment, 196, 214–5, 222–3
Singo–Indian rivalry, 11, 16, 24–5, 120, 141–2, 152, 158, 184, 187–91, 197–201, 205, 215, 218–20, 238, 241, 248

Index

Sino–Indian war (1962), 16, 20, 23, 104–5, 108, 110, 141, 144, 147, 152, 187, 189, 201, 211
Sino–Pakistan relations, 18, 106–7, 146, 181, 184–5, 187–99, 217–20
Sino–Soviet complex, 15, 17, 19–20, 158, 181–201
Sino–Soviet rivalry, 106, 181, 185–96, 201–2, 214–16, 227, 238–9
Socotra, 221
Somalia, 26, 221, 250
South Africa, 28, 29
Southern African complex, 28
South Asia, 6, 10, 13, 29, 127–31, 207
South Asian Regional Cooperation, 136
South Asian security complex, 10, 14–15, 19–25, 27, 28, 129, 135, 151, 158, 176, 195, 202, 207, 219–20, 235–52
Southeast Asian security complex, 9, 10, 14, 15, 16
Southeast Asian Treaty Organisation (SEATO), 182, 184, 193, 208, 211–13, 216
South Korea, 82, 171
Southwest Asia, 23, 31, 220, 247
South Yemen, 221
Soviet–American rivalry, 184, 222–4, 227–8, 238–9
Soviet Union, 7, 10, 11, 12, 15–16, 19–20, 24, 26, 30–1, 62, 73, 80–1, 101, 104–8, 112–13, 115, 119–20, 122, 130, 133, 135, 139, 148–53, 168–9, 173, 177, 181–202, 205–28, 237–9, 243–50
Speigel, S., 6
Sri Lanka, 10, 120, 128–30, 136–41, 153, 157–8, 226, 240, 242
Stalin, 183, 213
Status quo, 27–8, 235–40
Strategic Arms Limitation Talks (SALT), 225
Subrahmanyam, K., 10
Supercomplexes, 12
Superpower complex, 12, 17, 18–19, 205–28

Superpower rivalry, 120
Syria, 25, 29

Tamil Nadu, 42, 45, 46, 47, 138–9
Tamils, 108, 129–30, 136–40
Tamil United Liberation Front (TULF), 137–8
Tashkent Agreement, 112–13, 212, 214
Thailand, 171
Third World, 26, 30, 31, 159, 213, 224–5
Thompson, W., 6
Tibet, 130, 141–3, 146–7, 152, 157, 185, 188, 221, 242
Timor, 79
Tripura, 48
Turkey, 10, 26, 107, 170, 172, 209–10
Turks, 163

United Arab Emirates (UAE), 10, 162–7, 170–72, 175
United Arab Republic, 107; *see also* Egypt
United Nations, 100, 109, 112, 133, 135, 140–1, 144, 147, 152, 182, 185, 194–5
United Provinces, 67
United States, 11, 12, 13, 24, 26, 30–1, 83, 101–8, 112, 115–16, 120, 122, 136, 139, 149–51, 153, 165, 167–8, 170–3, 176, 181–7, 189, 192–5, 200, 205–28, 238–40, 242–3, 245–7, 251
Urdu, 39, 40, 74–5
Uttar Pradesh, 42, 143, 145

Vayrynen, R., 6
Vietnam, 15–16, 31, 139, 186, 198, 211, 214, 223, 248
Vishwa Hindu Parishad, 41

Weak states, 14, 60–1
West Bengal, 46, 47, 50, 145
Westminster, 70–71

Xiaoping, Deng, 197–8

Zedong, Mao, 184, 187, 197

RAYMOND H. FOGLER LIBRARY
DATE DUE